2023 Global Conference on Process Safety and Big Data

Frankfurt, Germany
24-25 October 2023

ISBN: 979-8-3313-3049-1

Printed from e-media with permission by:

Curran Associates, Inc.
57 Morehouse Lane
Red Hook, NY 12571

Some format issues inherent in the e-media version may also appear in this print version.

Printed with permission by Curran Associates, Inc. (2025)

For permission requests, please contact AIChE
at the address below.

AIChE
120 Wall Street, FL 23
New York, NY 10005-4020

Phone: (800) 242-4363
Fax: (203) 775-5177

www.aiche.org

Additional copies of this publication are available from:

Curran Associates, Inc.
57 Morehouse Lane
Red Hook, NY 12571 USA
Phone: 845-758-0400
Fax: 845-758-2633
Email: curran@proceedings.com
Web: www.proceedings.com

TABLE OF CONTENTS

empirisys

25 October 2023

Leading Process Safety in Digital Transformation

Gus Carroll

Contents

- My leadership journey
- How are we doing
- Why this matters
- How "big data" can help

empirisys

2

My Leadership Journey

UQUIFA, Agrate Brianza
Italy
May 2001

North Morecombe Terminal
UK
October 2017

empirisys

How Are We Doing?

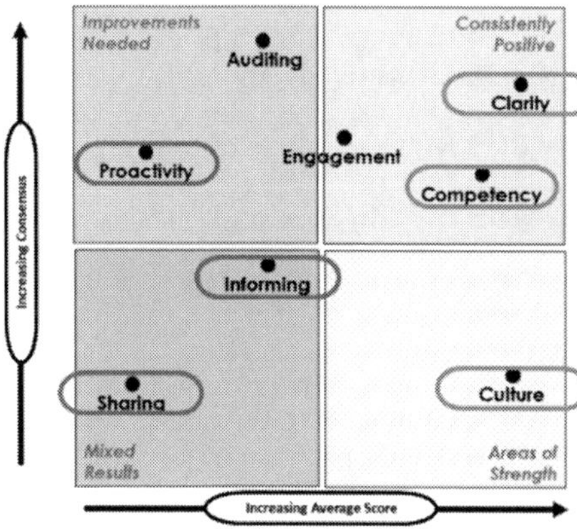

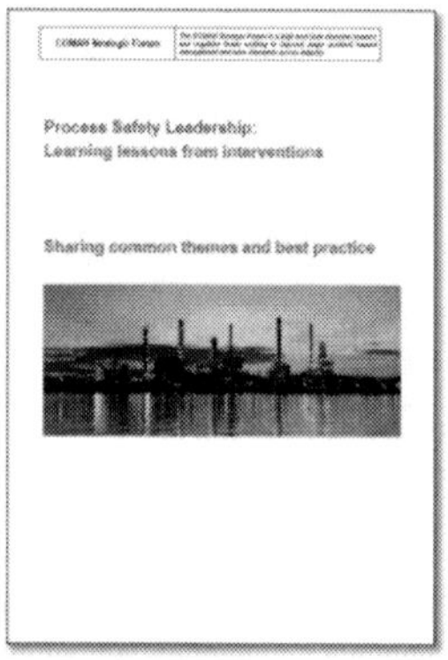

empirisys

[redacted]

Why This Matters

- Leaders set the tone of the organisation
- Critical to get this right to ensure
  - Bad news travels upwards without modification
  - Challenge and support are given equally
  - Individuals are not blamed

[redacted]

How "Big Data" Helps

- Data analytics & AI creates new insights from existing data
- PS experts make sense of the insights
- Leaders act on the insights

2023 GLOBAL CONFERENCE ON PROCESS SAFETY & BIG DATA

CCPS EPSC

2023 GLOBAL CONFERENCE ON PROCESS SAFETY & BIG DATA

CCPS EPSC

Leveraging Ex Equipment Inspections and Data Analysis for Sustainable Process Safety

Y. Mert Sönmez, Mehmet Dilaver
Chemical Engineer, MSc. *Chemical Engineer, MBA*

ProSCon Engineering - www.proscon.com.tr

ProSCon
pure process safety

Co-organized by CCPS & EPSC

2023 GLOBAL CONFERENCE ON PROCESS SAFETY & BIG DATA

CCPS EPSC

Y. Mert Sönmez

Chemical Engineer, MSc.

- ProSCon Engineering Co-Founder, Process Safety Consultant
- Process Safety Management and explosion protection process implementation consultancy and project coaching
- Consultancy for the design and implementation of risk management elements within the scope of process safety management
- One of the author team of the books Process Safety Management, Process Hazard Analyses and Explosion Protection Concept
- Member of the Chamber of Chemical Engineers in Union of Chambers of Turkish Engineers and Architects

Co-organized by CCPS & EPSC

2023 GLOBAL CONFERENCE ON PROCESS SAFETY & BIG DATA

CCPS EPSC

2023 GLOBAL CONFERENCE ON PROCESS SAFETY & BIG DATA

CCPS EPSC

OCTOBER 24 - 25, 2023
DECHEMA HOUSE | Frankfurt am Main, Germany

ATEX Sustainability Process

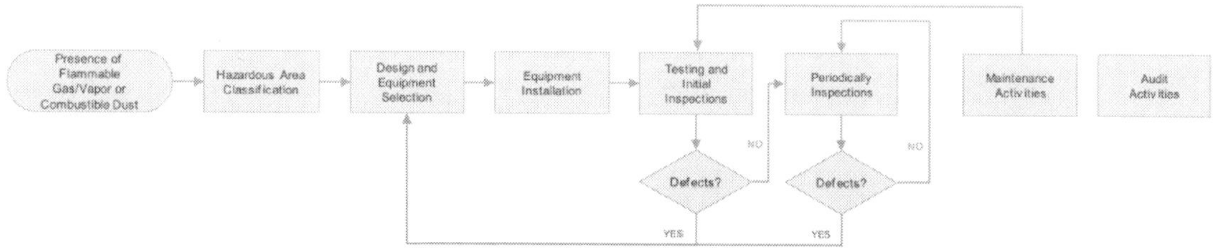

Co-organized by CCPS & EPSC

2023 GLOBAL CONFERENCE ON PROCESS SAFETY & BIG DATA

CCPS EPSC

OCTOBER 24 - 25, 2023
DECHEMA HOUSE | Frankfurt am Main, Germany

Inspection Process

IEC 60079-17

INTERNATIONAL STANDARD

NORME INTERNATIONALE

Regarding to IEC 60079-17 there are three grades of ex equipment inspections:

- Visual Inspection
- Close Inspection
- Detailled Inspection

Visual and close inspections can be performed with the equipment energized. No need to touch equipment, no need to electrical isolation and no need to production shutdown or stoppage.

Detailed inspections will generally require the equipment to be electrically isolated.

And also three types of ex equipment inspections:

- Initial Inspections
- Periodic Inspections
- Sample Inspections

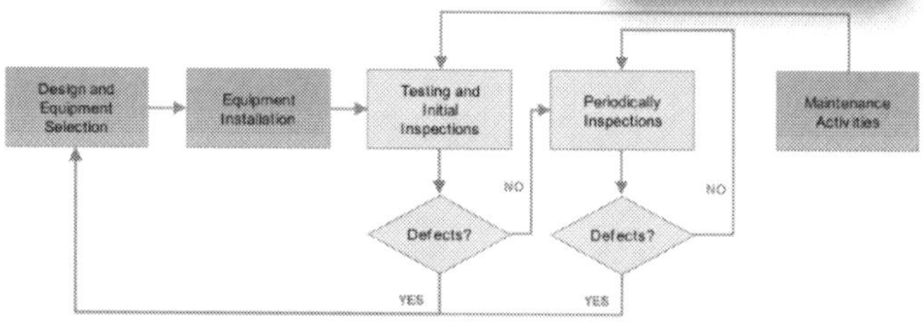

Co-organized by CCPS & EPSC

2023 GLOBAL CONFERENCE ON PROCESS SAFETY & BIG DATA

CCPS EPSC

OCTOBER 24 - 25, 2023
DECHEMA HOUSE | Frankfurt am Main, Germany

Inspection Process

Regarding to IEC 60079-17 there are three grades of ex equipment inspections:

- Visual Inspection
- Close Inspection
- Detailled Inspection

Mainly Investigate:

- Selection of equipment and components
- Installation or assembly problems
- General equipment condition
- Pre- and post- maintenance problems

And also three types of ex equipment inspections:

- Initial Inspections
- Periodic Inspections
- Sample Inspections

Period:

- Generally once per 3 years
- If needed combine with sample inspections for environmental effects (weather, corrosion, vibration, painting processes, dust and dirt, etc.)

Co-organized by CCPS & EPSC

2023 GLOBAL CONFERENCE ON PROCESS SAFETY & BIG DATA

CCPS EPSC

OCTOBER 24 - 25, 2023
DECHEMA HOUSE | Frankfurt am Main, Germany

Data Set of Study

Regarding to IEC 60079-17 there are three grades of ex equipment inspections:

- Visual Inspection
- Close Inspection
- Detailled Inspection

And also three types of ex equipment inspections:

- Initial Inspections
- Periodic Inspections
- Sample Inspections

35 facilities are inspected

Categorized by 4 industries

Total ATEX compliance ratio of data is 29,45%

Average plant life is 18 years

Up to 26,000 equipments are inspected

Each equipment has **132 datapoints**

Up to 3.4M data can be analizable

Co-organized by CCPS & EPSC

2023 GLOBAL CONFERENCE ON PROCESS SAFETY & BIG DATA

CCPS EPSC

OCTOBER 24 - 25, 2023
DECHEMA HOUSE | Frankfurt am Main, Germany

Inspection Process Problems

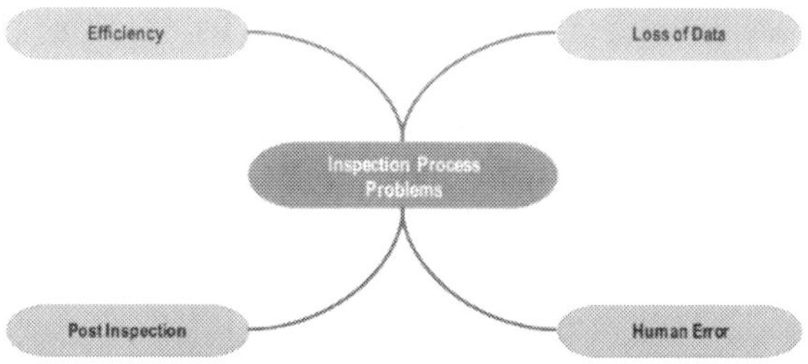

Co-organized by CCPS & EPSC

2023 GLOBAL CONFERENCE ON PROCESS SAFETY & BIG DATA

CCPS EPSC

OCTOBER 24 - 25, 2023
DECHEMA HOUSE | Frankfurt am Main, Germany

Inspection Process Problems

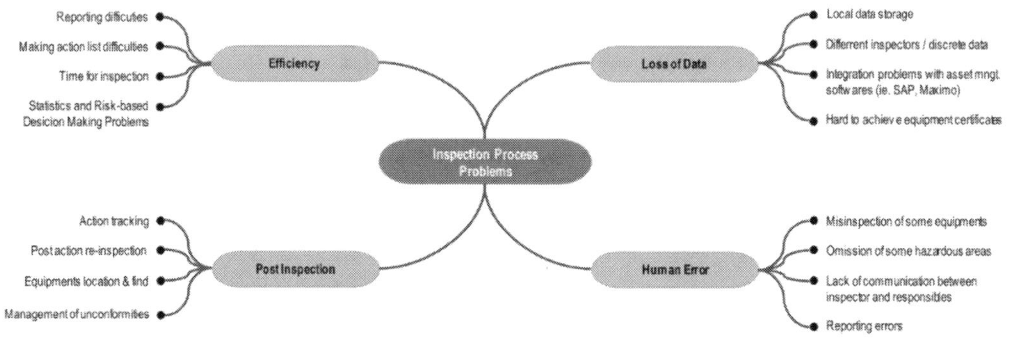

Co-organized by CCPS & EPSC

Tools and Method

ExA syst

- ExA zone
- ExA consequence
- ExA inspection
- ExA audit
- ExA competency

ExA inspection

- ExA inspection software is used in this study.
- Evaluations are made by data sets and error types/error classes.
- Chronic problems (gland selection, shroud usage, incorrect use of conduits, etc.) are identified.
- Faults are classified (design related, installation related, maintenance related, operation related, etc.)

Co-organized by CCPS & EPSC

2023 GLOBAL CONFERENCE ON PROCESS SAFETY & BIG DATA
CCPS EPSC
OCTOBER 24 - 25, 2023
DECHEMA HOUSE | Frankfurt am Main, Germany

Inspected Ex-equipments by Industries

(Industry; Number of inspected equipments; Percentage)

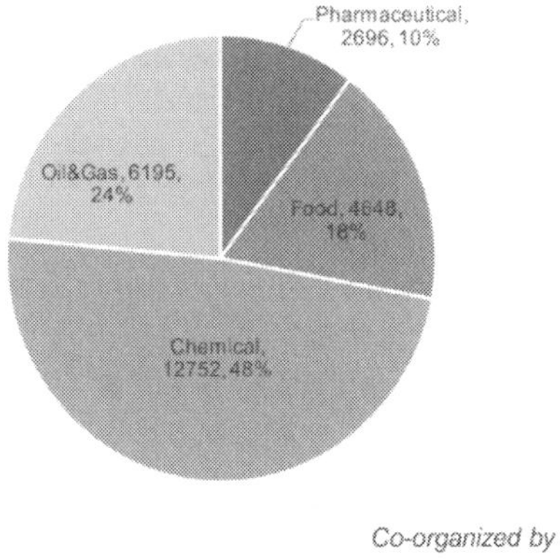

Total number of facilities are 35, ex-equipment compliance rate is 29.45%.

All data was selected to be from globally managed companies with facilities in different countries.

Facilities in **chemical industry** includes paint production, basic chemical raw materials production, chemical storage terminals and similar facilities.

Oil&Gas includes storage terminals and refineries.

Pharmaceutical includes API and final product facilities. By its nature, it also includes chemical processes such as solvent extraction, reaction and distillation, etc.

Food industry includes storage facilities, food production and manufacturers of intermediate products for food. By its nature, it is similar to pharmaceutical industry which have chemical processes.

Co-organized by CCPS & EPSC

2023 GLOBAL CONFERENCE ON PROCESS SAFETY & BIG DATA

CCPS EPSC

OCTOBER 24 - 25, 2023
DECHEMA HOUSE | Frankfurt am Main, Germany

Ex Equipment Conformity by Close Inspection
(IEC 60079-17)

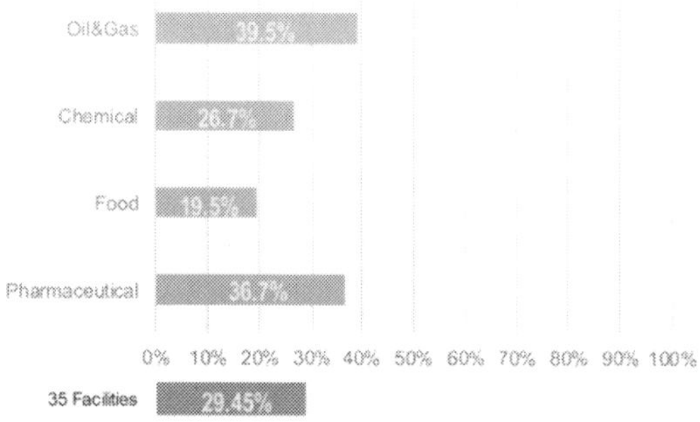

Oil&gas

High availability of "ex equipment".

External audits perform most likely from insurance companies, legal authorities and 3rd parties.

Availibility of standard / good engineering practices / industrial codes

Lack of awareness of such measures can cause some equipment to fail the ex-equipment inspection.

Incorrect use of shrouds or incorrect conduit system designs.

Co-organized by CCPS & EPSC

2023 GLOBAL CONFERENCE ON PROCESS SAFETY & BIG DATA

CCPS EPSC

OCTOBER 24 - 25, 2023
DECHEMA HOUSE | Frankfurt am Main, Germany

Ex Equipment Conformity by Close Inspection
(IEC 60079-17)

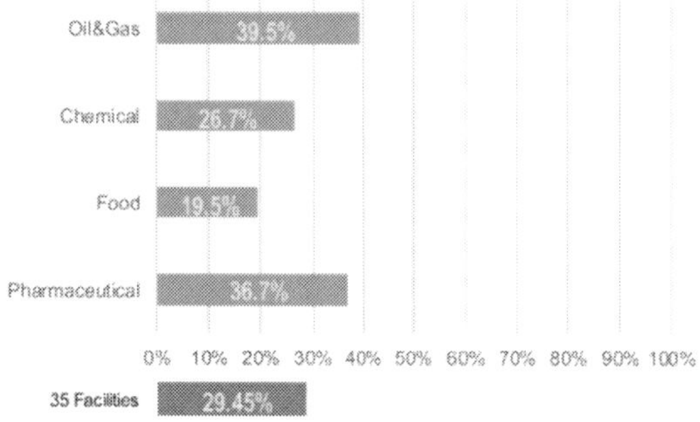

Chemical

High awareness about chemical hazards.

Incorrect hazardous area classification in chemical industry generally result in the use of incorrect ex-equipment.

Corrosion of chemicals is also critical to the condition of the equipment.

Co-organized by CCPS & EPSC

2023 GLOBAL CONFERENCE ON PROCESS SAFETY & BIG DATA

CCPS EPSC

OCTOBER 24 - 25, 2023
DECHEMA HOUSE | Frankfurt am Main, Germany

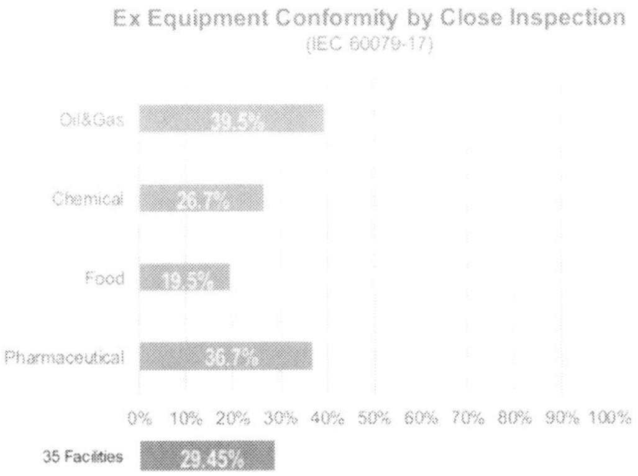

Food

Requirements for GMP application results with positive effect.

Dust leakage control and housekeeping practices are poorer than others. this allows dust to accumulate to form explosive dust clouds.

Co-organized by CCPS & EPSC

2023 GLOBAL CONFERENCE ON PROCESS SAFETY & BIG DATA

CCPS EPSC

OCTOBER 24 - 25, 2023
DECHEMA HOUSE | Frankfurt am Main, Germany

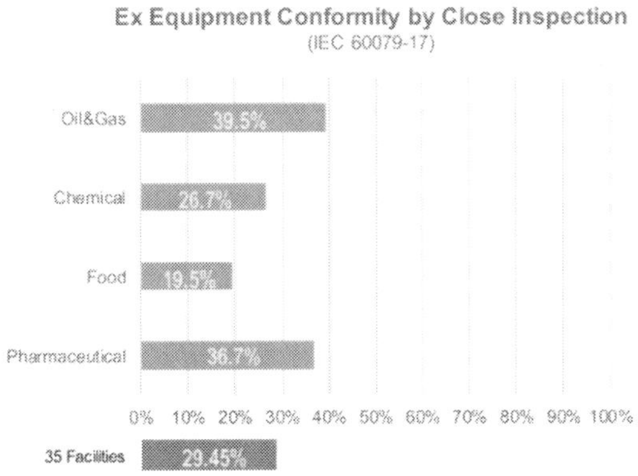

Pharmaceutical

Requirements for GMP application results with positive effect.

Design and installation verification processes have a positive impact.

The painting application required by GMP in the pharmaceutical industry has a negative impact on ex equipment.

Co-organized by CCPS & EPSC

In this study all data is classified according to the following protection types.

Flame proof Enclosure **Type d/tD**	Increased Safety **Type e**	Intrinsic Safety **Type i**	Non Indencive **Type n**	Others Protection Types

2023 GLOBAL CONFERENCE ON PROCESS SAFETY & BIG DATA
CCPS EPSC
OCTOBER 24 - 25, 2023
DECHEMA HOUSE | Frankfurt am Main, Germany

Ex Equipment Conformity by Protection Types

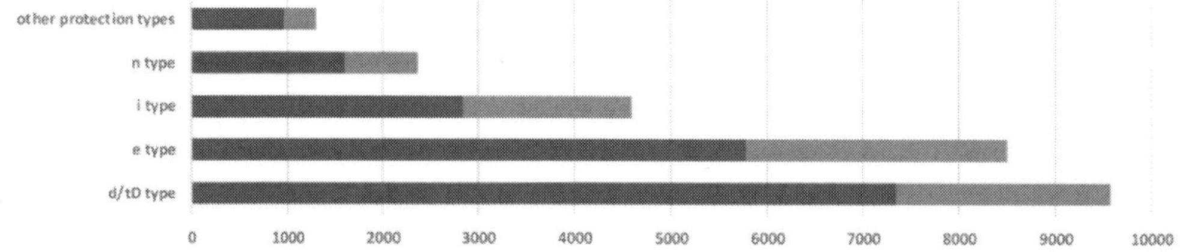

	d/tD type	e type	i type	n type	other protection types
Total Number of Failed Equipments	7349	5782	2845	1613	960
Total Number of Passed Equipments	2216	2707	1742	741	336
Conformity%	23.17%	31.89%	37.98%	31.48%	25.93%

Co-organized by CCPS & EPSC

2023 GLOBAL CONFERENCE ON PROCESS SAFETY & BIG DATA

CCPS EPSC

OCTOBER 24 - 25, 2023
DECHEMA HOUSE | Frankfurt am Main, Germany

Ex Equipment Findings by Protection Types and Type of Finding

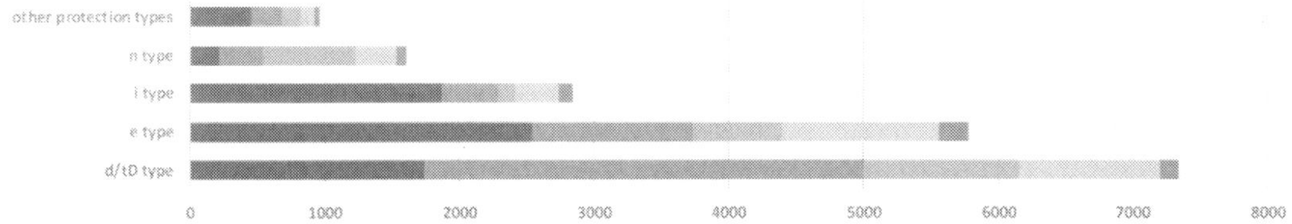

	d/tD type	e type	i type	n type	other protection types
Design related problems	1745	2545	1874	214	451
Assembly related problems	3254	1198	416	329	231
Maintenance related problems	1163	652	122	685	143
Operation related problems	1045	1168	326	304	95
Other	142	219	107	81	40

Co-organized by CCPS & EPSC

2023 GLOBAL CONFERENCE ON PROCESS SAFETY & BIG DATA

CCPS EPSC

OCTOBER 24 - 25, 2023
DECHEMA HOUSE | Frankfurt am Main, Germany

d/tD type Protection

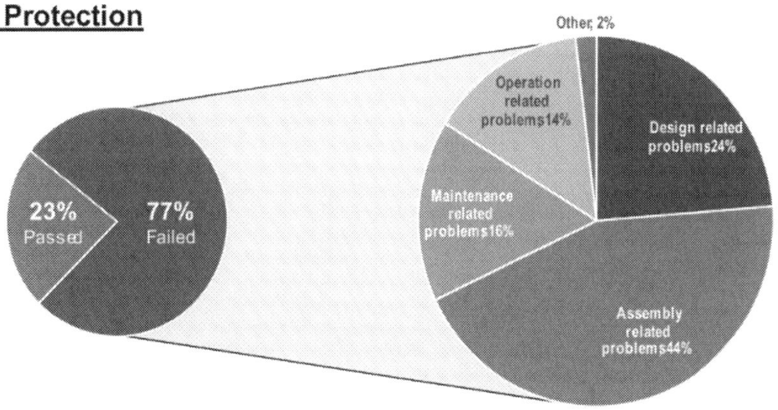

Co-organized by CCPS & EPSC

2023 GLOBAL CONFERENCE ON PROCESS SAFETY & BIG DATA

CCPS EPSC

OCTOBER 24 - 25, 2023
DECHEMA HOUSE | Frankfurt am Main, Germany

d/tD type Protection

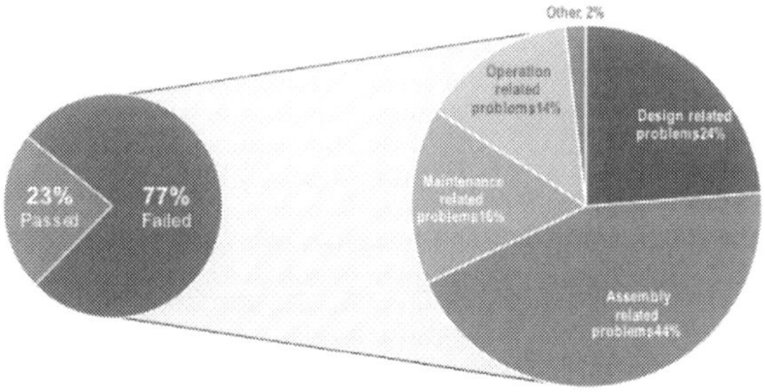

- The largest number of non-conformities is related to installation with 44%.
- Incompatibilities of connectors such as glands and blind plugs are at the top of installation errors.
- Painting of equipment blocks flamepaths and causes the maximum permissible paint thickness on the equipment to be exceeded
- Lack of bolts is common after the equipment has been opened for maintenance.

Co-organized by CCPS & EPSC

2023 GLOBAL CONFERENCE ON PROCESS SAFETY & BIG DATA

CCPS EPSC

OCTOBER 24 - 25, 2023
DECHEMA HOUSE | Frankfurt am Main, Germany

e type Protection

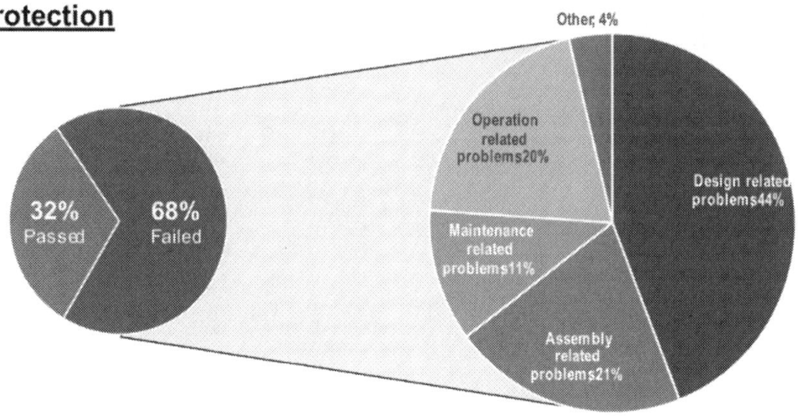

Co-organized by CCPS & EPSC

2023 GLOBAL CONFERENCE ON PROCESS SAFETY & BIG DATA

CCPS EPSC

OCTOBER 24 - 25, 2023
DECHEMA HOUSE | Frankfurt am Main, Germany

e type Protection

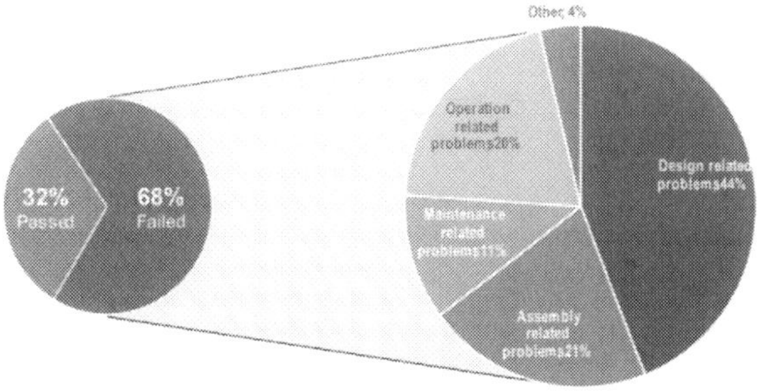

- The largest number of non-conformities is related to design with 44%.
- There is a high rate of terminal design errors to be placed inside the equipment.
- Incompatibilities of connectors such as glands and blind plugs are at the top of installation errors.
- Lack of bolts is common after the equipment has been opened for maintenance.

Co-organized by CCPS & EPSC

2023 GLOBAL CONFERENCE ON PROCESS SAFETY & BIG DATA

CCPS EPSC

OCTOBER 24 - 25, 2023
DECHEMA HOUSE | Frankfurt am Main, Germany

i type Protection

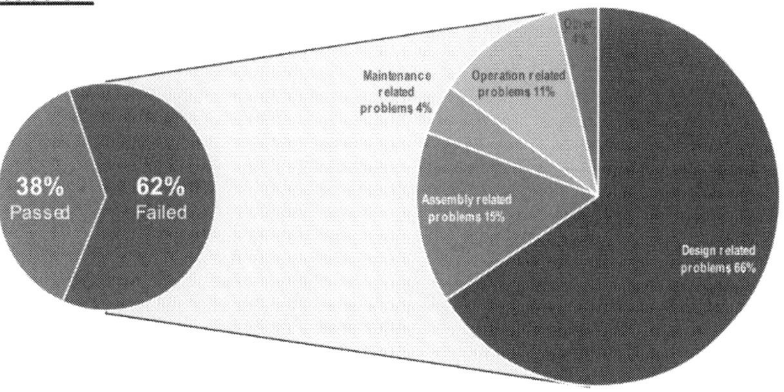

Co-organized by CCPS & EPSC

2023 GLOBAL CONFERENCE ON PROCESS SAFETY & BIG DATA

CCPS EPSC

OCTOBER 24 - 25, 2023
DECHEMA HOUSE | Frankfurt am Main, Germany

i type Protection

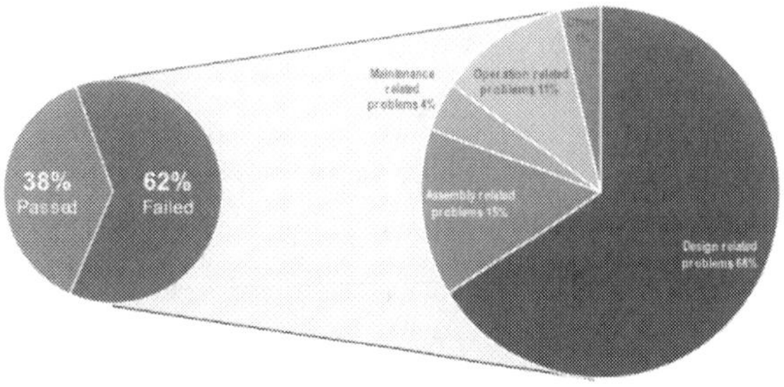

The largest number of non-conformities is related to design with 66%.

The basic design problem is that the IS circuit is not installed, not connected to a barrier or incorrect barrier selection.

Co-organized by CCPS & EPSC

2023 GLOBAL CONFERENCE ON PROCESS SAFETY & BIG DATA

CCPS EPSC

OCTOBER 24 - 25, 2023
DECHEMA HOUSE | Frankfurt am Main, Germany

n type Protection

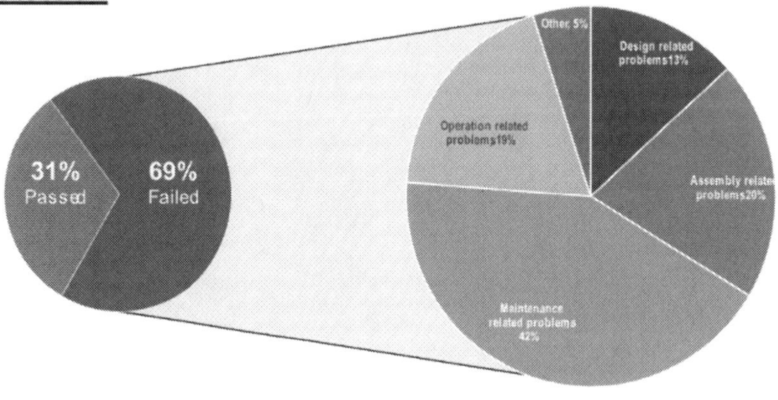

Co-organized by CCPS & EPSC

n type Protection

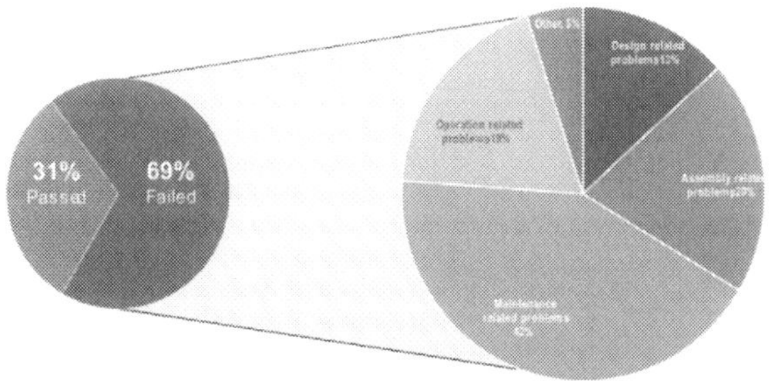

The largest number of non-conformities is related to maintenance with 42%.

It has been observed that problems increase especially in engines after maintenance. For example, ventilation fan problems, overheating problems, thermistor faults etc.

Incompatibilities of connectors such as glands and blind plugs are at the top of installation errors.

Lack of bolts is common after the equipment has been opened for maintenance.

2023 GLOBAL CONFERENCE ON PROCESS SAFETY & BIG DATA
CCPS EPSC
OCTOBER 24 - 25, 2023
DECHEMA HOUSE | Frankfurt am Main, Germany

energy INSTITUTE

Guidelines for managing inspection of Ex electrical equipment ignition risk in support of IEC 60079-17

Ex-RBI

Ex – Risk Based Inspection

- Requires categorisation of non-conformities after periodic inspection (Safety, Integrity & Housekeeping)
- Prioritisation is carried out together with the categorisation.
- In this method, following can be determined:
  - The scope and period of sample inspection
  - Action prioritisation and downtime/shutdown possibilities

Co-organized by CCPS & EPSC

2023 GLOBAL CONFERENCE ON PROCESS SAFETY & BIG DATA

CCPS EPSC

OCTOBER 24 - 25, 2023
DECHEMA HOUSE | Frankfurt am Main, Germany

	Fault Description	Severity	Priority	# of Actions
Internal inspections (detailed):				
Z	Is Zener barrier earth bond <1 ohm	Safety	1	986
Q	Inadequate segregation IS and pow er	Safety	1	154
T	Ex 'e' loose terminations	Safety	1	-
N	Ex 'e' unconnected cores	Integrity	2	-
W	Ex 'e' terminations, insulation degradation (w ater ingress)	Safety	1	-
F	Ex 'd' flame path damaged	Safety	1	-
F	Ex 'd' flame path impeded	Integrity	2	206
P	Ex 'p' pressurisation control system fault	Safety	1	-
X	Ex 'p' certification special requirements X	Safety	1	-
External inspections (visual and close):				
B	Covers/bolts	Safety	1	3210
G	Glands/stop ends	Integrity	2	11542
E	Earthing	Integrity	2	9851
U	Uncertified equipment used	Safety	1	10858
H	Certificate inadequate	Integrity	3	1052
S	Fed-from labels	Safety	1	3712
I	Ingress	Housekeeping	4	741
D	Damage	Housekeeping	4	1365
M	Mounting	Housekeeping	4	1259
C	Corrosion	Housekeeping	4	5426
R	Redundant (Unnecessary)	Housekeeping	5	89
L	Duty/tag numbers	Housekeeping	5	7816

energy INSTITUTE

Safety: priority 1 faults where corrective action should be taken in the short term (e.g. < one week).
Integrity: priority 2-3 faults where corrective action should be taken in the medium term (e.g. ≤ three months).
Housekeeping: priority 4-5 faults that do not affect level of risk but represent a non-compliance with standards (e.g. missing or damaged labelling) where action should be taken in the long term (e.g. ≤12 months such as during the next planned maintenance).

Co-organized by CCPS & EPSC

2023 GLOBAL CONFERENCE ON PROCESS SAFETY & BIG DATA

CCPS EPSC

OCTOBER 24 - 25, 2023
DECHEMA HOUSE | Frankfurt am Main, Germany

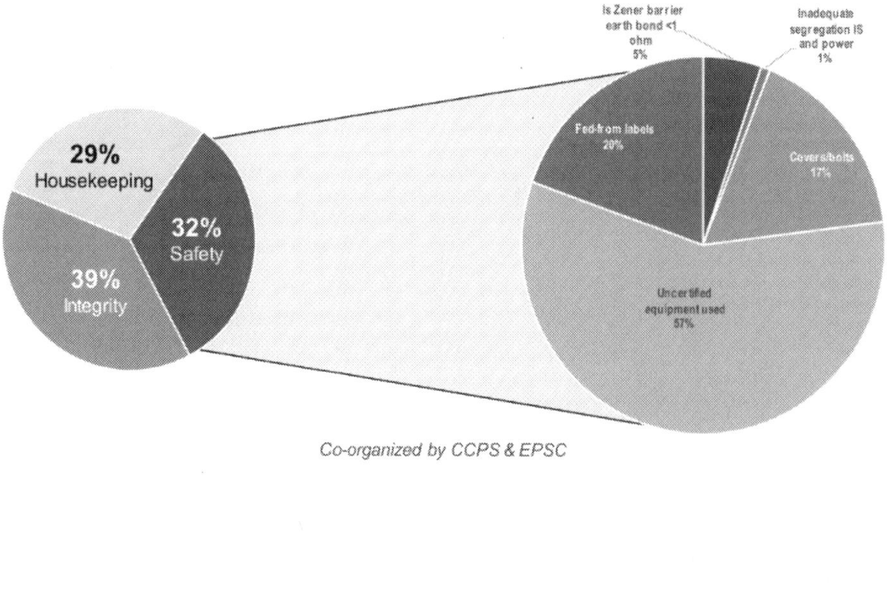

Co-organized by CCPS & EPSC

2023 GLOBAL CONFERENCE ON PROCESS SAFETY & BIG DATA

CCPS EPSC

OCTOBER 24 - 25, 2023
DECHEMA HOUSE | Frankfurt am Main, Germany

Recommendations

Industry Based Recommendations

- In the pharmaceutical industry, it is recommended to plan sample inspection after painting applications of equipments in accordance with GMP requirement.
- In the food industry, it is recommended to determine the amount of dust, housekeeping planning and operate consistently in order to control dust and prevent dust leaks.
- In the chemical industry, the importance of HAC studies should be emphasised.
- In oil & gas, it is recommended to increase the competence of personnel to protect the equipment from external factors.

Co-organized by CCPS & EPSC

2023 GLOBAL CONFERENCE ON PROCESS SAFETY & BIG DATA

CCPS EPSC

OCTOBER 24 - 25, 2023
DECHEMA HOUSE | Frankfurt am Main, Germany

Recommendations

Equipment Based Recommendations

- Design related problems are mostly seen in e and i types. It is recommended that the terminal design and IS circuit design of this type of equipment be carried out by competent persons.
- Installation related problems are mostly seen in type d. It is recommended to perform initial inspection after installation.
- Maintenance related problems are mostly seen in type n. It is recommended to perform pre-startup control and inspection after maintenance.
- Operation related problems are mostly seen in types n and e. It is recommended to plan sample inspections during the inspection plan.

Co-organized by CCPS & EPSC

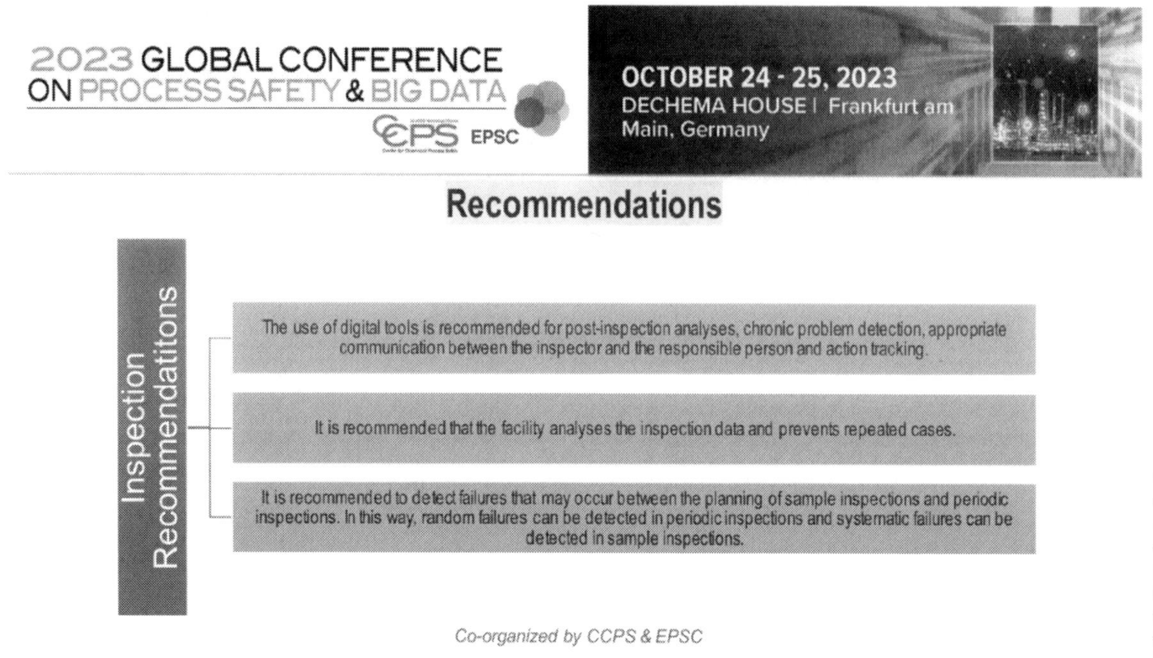
2023 GLOBAL CONFERENCE ON PROCESS SAFETY & BIG DATA
CCPS EPSC
OCTOBER 24 - 25, 2023
DECHEMA HOUSE | Frankfurt am Main, Germany
Recommendations
Inspection Recommendatitons
The use of digital tools is recommended for post-inspection analyses, chronic problem detection, appropriate communication between the inspector and the responsible person and action tracking.
It is recommended that the facility analyses the inspection data and prevents repeated cases.
It is recommended to detect failures that may occur between the planning of sample inspections and periodic inspections. In this way, random failures can be detected in periodic inspections and systematic failures can be detected in sample inspections.
Co-organized by CCPS & EPSC

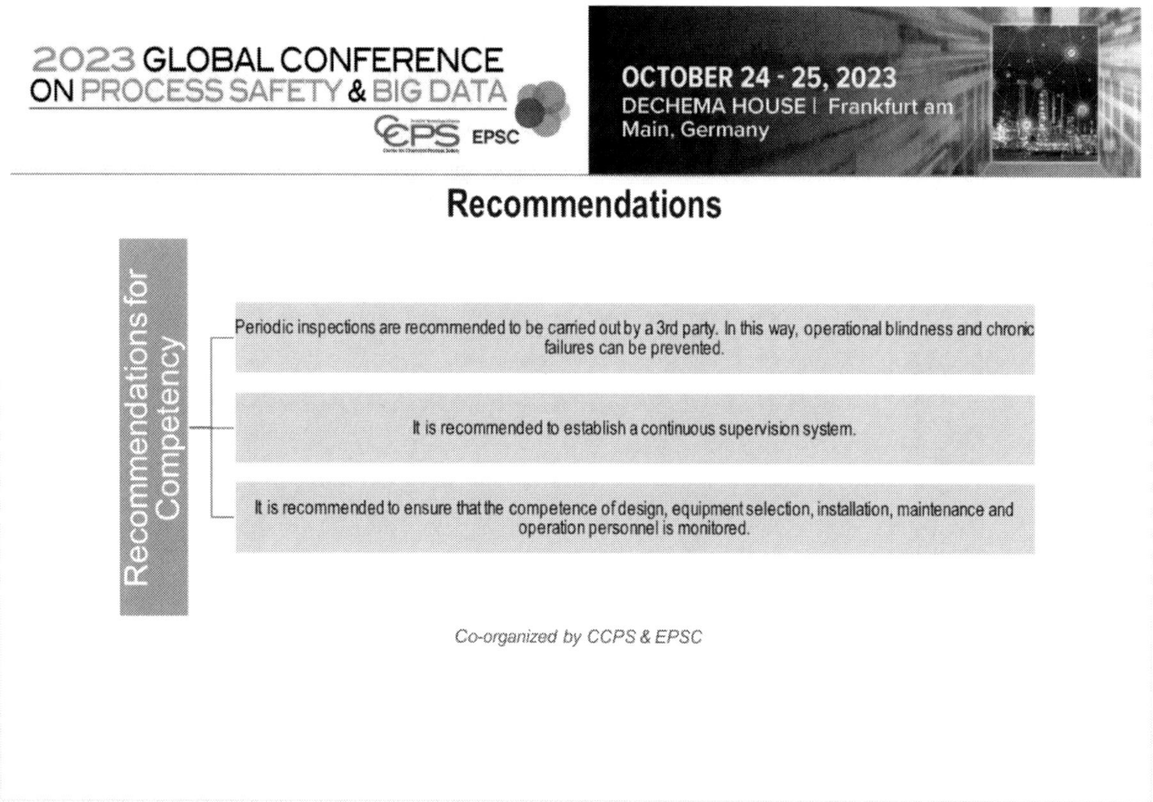
2023 GLOBAL CONFERENCE ON PROCESS SAFETY & BIG DATA
CCPS EPSC
OCTOBER 24 - 25, 2023
DECHEMA HOUSE | Frankfurt am Main, Germany
Recommendations
Recommendations for Competency
Periodic inspections are recommended to be carried out by a 3rd party. In this way, operational blindness and chronic failures can be prevented.
It is recommended to establish a continuous supervision system.
It is recommended to ensure that the competence of design, equipment selection, installation, maintenance and operation personnel is monitored.
Co-organized by CCPS & EPSC

2023 GLOBAL CONFERENCE ON PROCESS SAFETY & BIG DATA

CCPS EPSC

OCTOBER 24 - 25, 2023
DECHEMA HOUSE | Frankfurt am Main, Germany

Recommendations

Organizational Recommandations

- If action planning will be done with Ex-RBI, it is recommended that sample inspection planning in risk based approach should be selected from equipment with operation related problems.
- The situation "Uncertified equipment used" is a serious safety issue and not possible within the time recommended by energy institute (<1 week).
- It is recommended that the verification dossier is complete and as-built.
- An integrated system including elements of PSM is recommended (especially with elements of asset integrity, critical equipment management, management of change, PSSR, competence, safety culture, PHA, training, audit and metrics)

Co-organized by CCPS & EPSC

2023 GLOBAL CONFERENCE ON PROCESS SAFETY & BIG DATA

CCPS EPSC

OCTOBER 24 - 25, 2023
DECHEMA HOUSE | Frankfurt am Main, Germany

Thank you for your attention

Y. Mert Sönmez
Chemical Engineer, MSc.
Co-Founder

mert.sonmez@proscon.com.tr

ProSCon
pure process safety

ProSCon Engineering - www.proscon.com.tr

2023 GLOBAL CONFERENCE ON PROCESS SAFETY & BIG DATA

CCPS EPSC

2023 GLOBAL CONFERENCE ON PROCESS SAFETY & BIG DATA

CCPS EPSC

Advanced Barrier Health Monitoring

A Process Safety Analytics Case Study

Presenter : Jerson Wattie

Company: The National Gas Company Of Trinidad and Tobago Limited

2023 GLOBAL CONFERENCE ON PROCESS SAFETY & BIG DATA

CCPS EPSC

OCTOBER 24 - 25, 2023
DECHEMA HOUSE | Frankfurt am Main, Germany

DISCLAIMER

The National Gas Company of Trinidad and Tobago Limited (NGC) has agreed to make this presentation to the attendees of this event. NGC makes no representation as to the accuracy of the information contained herein and advises that no reliance should be placed on that information. The information in this presentation is based primarily on publicly available sources, which NGC is not responsible for preparing or verifying.

The information provided shall not be binding on NGC, its shareholders, board of directors, employees and/or agents, and has no legal effect and in no circumstances shall the contents be relied upon for any reason whatsoever. No responsibility is accepted by or on behalf of NGC, its shareholders, board of directors, employees and/or agents for any decisions made by the recipients of such information pursuant to receipt of the information nor shall NGC be responsible for any direct or indirect consequences as a result of any reliance on the information being provided by NGC herein.

2023 GLOBAL CONFERENCE ON PROCESS SAFETY & BIG DATA

CCPS EPSC

OCTOBER 24 - 25, 2023
DECHEMA HOUSE | Frankfurt am Main, Germany

CONTEXT

1. Our business is managing risk
2. We use barrier modeling to help visualize risks
3. Data converts information to make decisions about those risks
4. Data is becoming more complex, and operators are being called to make decisions
5. Getting it right saves lives

Co-organized by CCPS & EPSC

Subject to disclaimer

2023 GLOBAL CONFERENCE
ON PROCESS SAFETY & BIG DATA
CCPS EPSC
OCTOBER 24 - 25, 2023
DECHEMA HOUSE I Frankfurt am Main, Germany
We are the NGC Group
The NGC Group of Companies, moving towards our vision as a recognized global leader in the development of sustainable energy related businesses
Technical Advisory Services
Equity Participation/JV
Clean Energy & CNG
Integrated along Nat. Gas Value Chain
LNG & Commodity Trading
Merchant & Aggregator
Port, Marine & Estate Infrastructure
Gas Processing
Pipeline Infrastructure
National Energy
Co-organized by CCPS & EPSC
Subject to disclaimer

2023 GLOBAL CONFERENCE
ON PROCESS SAFETY & BIG DATA
CCPS EPSC
OCTOBER 24 - 25, 2023
DECHEMA HOUSE I Frankfurt am Main, Germany
Where we going
Systematic and in Control of Significant Risks. Focus barrier health data.
Where we are
Managing risk using Barrier model. Focus visualizing risk
Not effectively managing risk. Focus on paper
Where we were
Co-organized by CCPS & EPSC
Subject to disclaimer

2023 GLOBAL CONFERENCE ON PROCESS SAFETY & BIG DATA

CCPS EPSC

OCTOBER 24 - 25, 2023
DECHEMA HOUSE | Frankfurt am Main, Germany

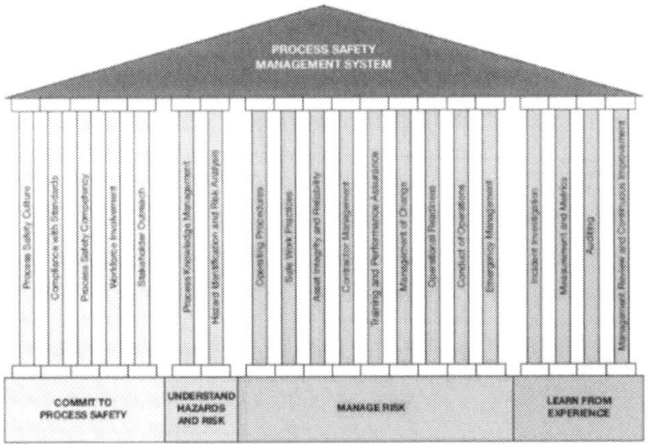

Co-organized by CCPS & EPSC

Subject to disclaimer

2023 GLOBAL CONFERENCE ON PROCESS SAFETY & BIG DATA

CCPS EPSC

OCTOBER 24 - 25, 2023
DECHEMA HOUSE | Frankfurt am Main, Germany

Future State: Architecture

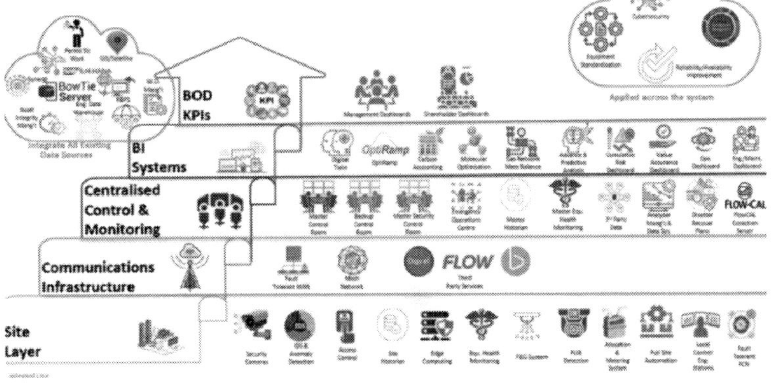

Co-organized by CCPS & EPSC

Subject to disclaimer

2023 GLOBAL CONFERENCE
ON PROCESS SAFETY & BIG DATA

CCPS EPSC

OCTOBER 24 - 25, 2023
DECHEMA HOUSE | Frankfurt am Main, Germany

IMPLEMENTATION INSIGHTS

- Decision to use Bow Tie Server as a fully integrated solution
- Training cross functional team
- Back pass using previous incidents to analyze
- Sharing Lessons

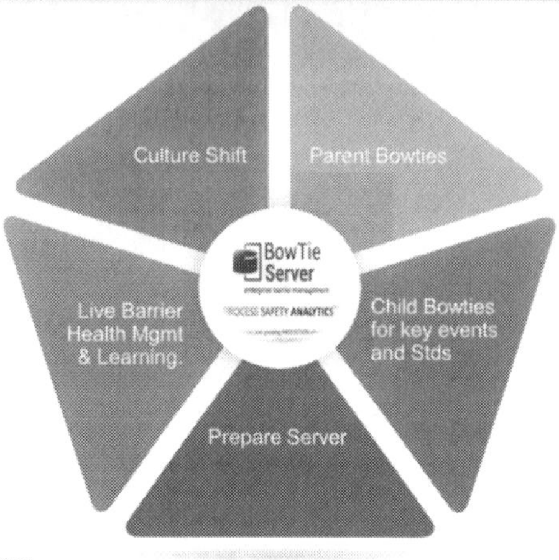

Co-organized by CCPS & EPSC

Subject to disclaimer

IMPLEMENTATION INSIGHTS

2023 GLOBAL CONFERENCE
ON PROCESS SAFETY & BIG DATA

CCPS EPSC

OCTOBER 24 - 25, 2023
DECHEMA HOUSE | Frankfurt am Main, Germany

CUMULATIVE RISK MODELLING

Depth Awareness

1. Develop a learning focus. Questioning..
2. Boundaries- Monitoring operating limits
3. Verifying- recording area details, multiple times/day

Cumulative Risk Modelling

The tool draws data from source e.g. Permit to Work and (CMMS) and combines the data to provide a traffic lighted risk profile for health of installations.

+	−
1) Improves layers of protection insights	1) More Data, More Complications
2) Real Time Reporting	2) Data Entry Quality
3) Predictive Analysis	3) Human Factors- Rule breaking, bias
4)Optimization	

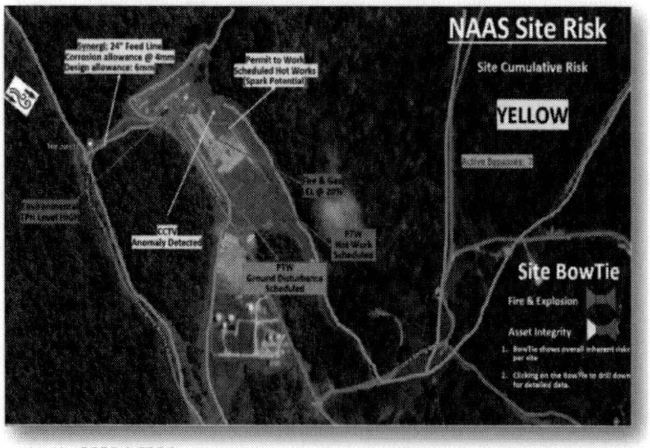

Co-organized by CCPS & EPSC

Subject to disclaimer

2023 GLOBAL CONFERENCE ON PROCESS SAFETY & BIG DATA

CCPS EPSC

OCTOBER 24 - 25, 2023
DECHEMA HOUSE | Frankfurt am Main, Germany

INCIDENT CASE STUDY

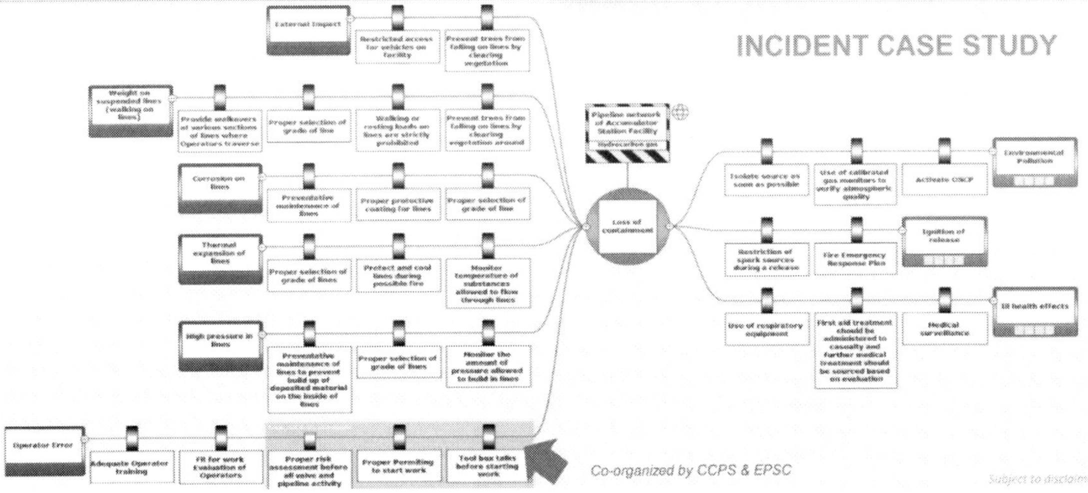

Co-organized by CCPS & EPSC

Subject to disclaimer

2023 GLOBAL CONFERENCE ON PROCESS SAFETY & BIG DATA

CCPS EPSC

OCTOBER 24 - 25, 2023
DECHEMA HOUSE | Frankfurt am Main, Germany

INCIDENT CASE STUDY:

Causation Branch

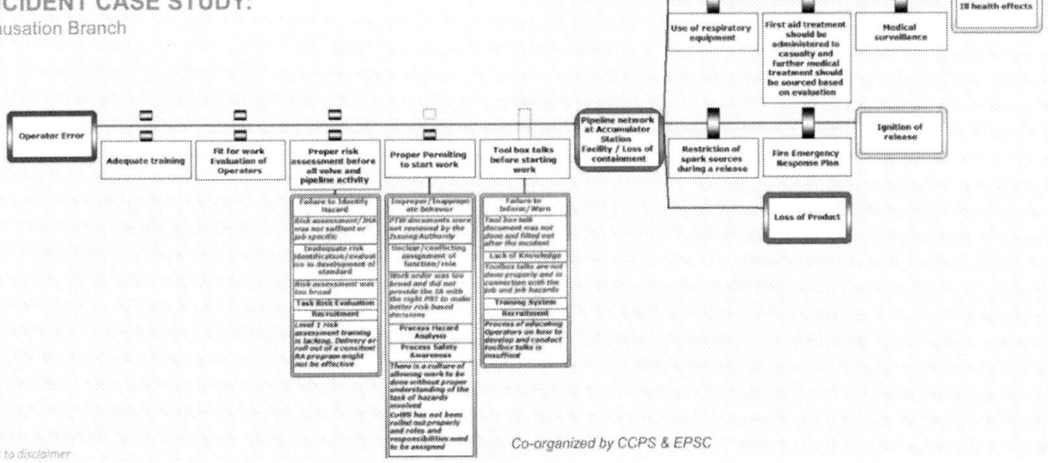

Co-organized by CCPS & EPSC

Subject to disclaimer

2023 GLOBAL CONFERENCE ON PROCESS SAFETY & BIG DATA

CCPS EPSC

OCTOBER 24 - 25, 2023
DECHEMA HOUSE | Frankfurt am Main, Germany

INCIDENT CASE STUDY:
Using Audit Function to Visualize Barrier Health

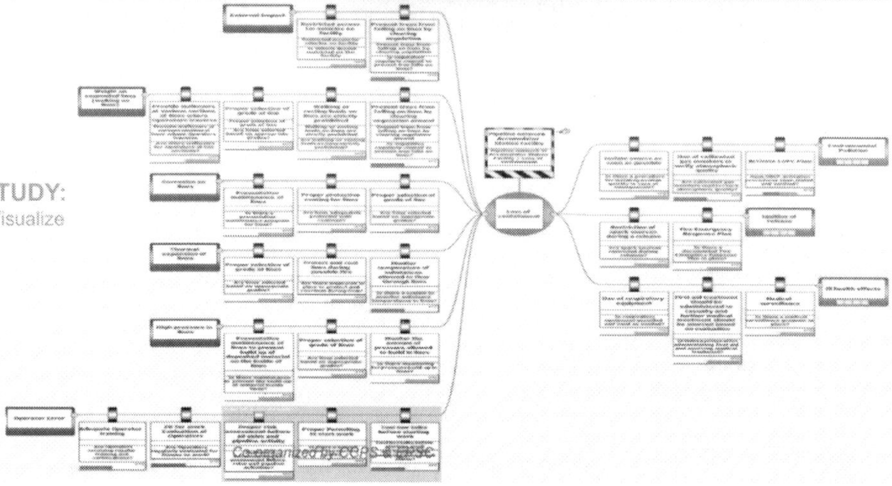

Co-organized by CCPS & EPSC

Subject to disclaimer

2023 GLOBAL CONFERENCE ON PROCESS SAFETY & BIG DATA

CCPS EPSC

OCTOBER 24 - 25, 2023
DECHEMA HOUSE | Frankfurt am Main, Germany

Better Analysis & Machine Learning

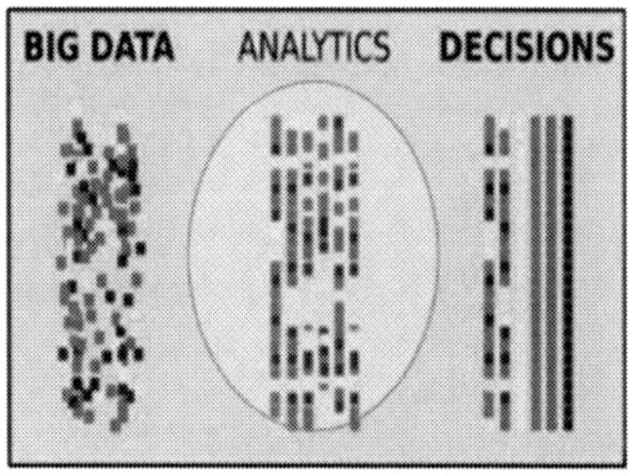

OPPORTUNITIES
- Key to better analysis & learning is in accuracy.
  - Human Factors in investigating
    - Making assumptions, false trails, narrowness (Energy Institute, 2008)

ML POTENTIAL
- Magnify expert investigator skills
- Generate prescriptive data analysis
- Increase transparency & optimization

Co-organized by CCPS & EPSC

Subject to disclaimer

2023 GLOBAL CONFERENCE ON PROCESS SAFETY & BIG DATA

CCPS EPSC

OCTOBER 24 - 25, 2023
DECHEMA HOUSE | Frankfurt am Main, Germany

WRAP UP

1. Manage Operational Risk
2. Visualize risk using barrier modelling
3. Reliable data to make decisions prevents us from getting it wrong

Co-organized by CCPS & EPSC

Big Data Analytics Potential in Unifying Digital Safety Solutions

Lama Aldosary
10/24/2023

SafeLife - *One-stop Shop Safety Digital Platform*

Digital safety platform that enables Aramco organizations and contractors to manage their safety while providing oversight and monitoring capabilities

2

Saudi Aramco: Public

Background & Evolution

Saudi Aramco: Public

3

System Expansion

- Employee Incident Management
- Safety Recommendations System
- Contractor Incident Management
- Inspection Management
- Corporate Safety Dashboard
- Risk Management
- AI Risk Prediction
- SafeTravel
- Safety Chatbot
- Contractor Safety Indices
- Safety Courses Tracking
- Safety Management System
- Fire Protection Incident
- Management of Change
- More...

Saudi Aramco: Public

4

Inspection Management

- Compliance Review
- Quarterly Safety Inspections
- Field Safety Inspections
  - Facility Compliance Team
  - Corporate Safety Team
  - Contractor Safety Team

5

Saudi Aramco: Public

Dashboards & Reports

- Dynamic and all-encompassing
- Suitable for all authorization levels
- Dashboard adaptability depending on module need

Saudi Aramco: Public

6

Emergency Preparedness

Users have the capability to arrange drills, oversee their schedules, and generate drill reports to effectively administer emergency preparedness efforts.

7

Saudi Aramco: Public

Emergency Preparedness

Users have the capability to arrange drills, oversee their schedules, and generate drill reports to effectively administer emergency preparedness efforts.

8

Saudi Aramco: Public

Safe Travel

Users can Schedule Journeys on the mobile application

911 button for instant alerts and immediate search and rescue

A Comprehensive dashboard for supervisors and journey coordinators

Saudi Aramco: Public

9

Safety Chatbot

Avails information to users at any time, summarizes informative feedback, and offers essential hyperlinks to relevant governing documentation.

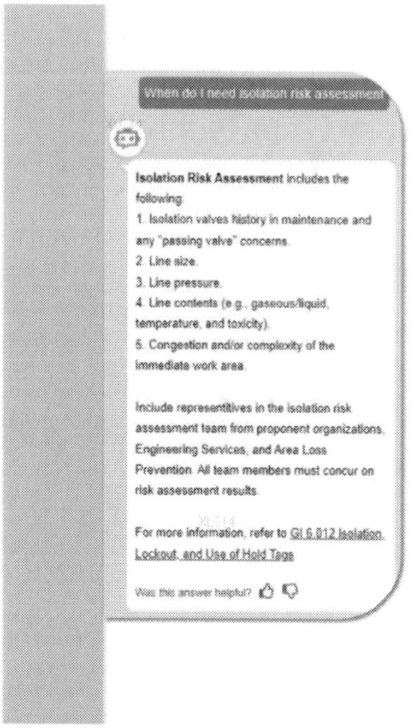

10

Saudi Aramco: Public

Risk Management

- Bowties
- PrHA
- HAZOP

Risk Prediction

A model trained on SafeLife data to list the risks associated with each activity and provide recommendations to address them

11

Saudi Aramco: Public

Risk Management

- Bowties
- PrHA
- HAZOP

Risk Prediction

A model trained on SafeLife data to list the risks associated with each activity and provide recommendations to address them

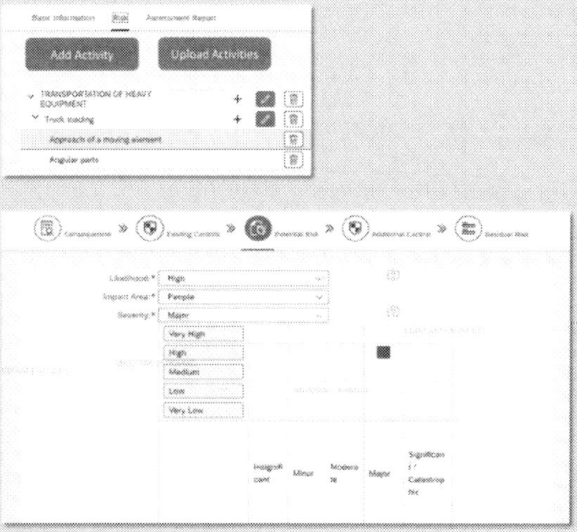

12

Saudi Aramco: Public

Safety Management System Module

Formulating KPIs using SafeLife variables

Managing safety expectations and deliverables

Conducting gap analysis

Safety management tracking

13

Saudi Aramco: Public

Lessons Learned

Create versatile modules that allow easy addition of new types

Equip users with flexibility of creation to show you what they need

Anticipate data links by evaluating modules within full system

Unifying safety systems will advance growth

14

Saudi Aramco: Public

أرامكو السعودية
saudi aramco

Hovione

In it for life

Data Analysis to Prevent Dust Explosion

Elsa Gonçalves
2023.10.24

Agenda

- **Dust Hazards**
- **Hovione classical approach**
- **Inherently Safe – work on the prevention**

Dust Hazards

- Dust explosions accidents occur when fine particles of dust are ignited
- Dust materials such as rubber and plastic products, organic powders, primary metal, lumber, wood products, food products, among others, have high potential to ignite
- Dust hazards affects different industries such as chemical and pharmaceuticals

CSB Reports Chemical Dust Explosions are a "Serious Problem" - Nearly 200 Accidents took 100 Lives; Data, Testimony Noted at Public Hearing Today

June 22, 2005

Washington, DC, June 22, 2005 - Carolyn Merritt, Chairman of the U.S. Chemical Safety and Hazard Investigation Board (CSB) declared chemical dust explosions in the United States are a "serious industrial safety problem" in opening a day-long hearing into the hazards of such accidents. Chairman Merritt noted the CSB's preliminary research reveals that nearly 200 dust fires and explosions have occurred in U.S. industrial facilities over the past 25 years, resulting in approximately 100 fatalities and 600 injuries.

"Dust explosions are preventable", she said at the hearing, convened by the CSB as part of an ongoing investigation into the causes and prevention of dust explosions, like the tragedies in Kinston, North Carolina (West Pharmaceutical Services), Corbin, Kentucky (CTA Acoustics), and Huntington, Indiana (Hayes Lemmerz), in 2003. A total of 14 people were killed and 81 injured in these three accidents. Information on the accidents can be found at www.csb.gov.

Hovione classical approach

- Test dust materials such as SRM, APIs, excipients
- Safety Database (SaDa) in 2022
- 640 dust reports issued in 10 years
- 97% led to the same safeguards

No more dust tests **campaign**

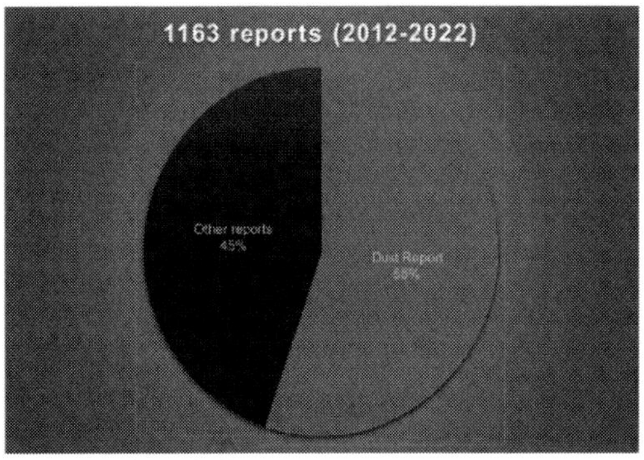

Extracted from SaDa

Minimum Ignition Energy (MIE)

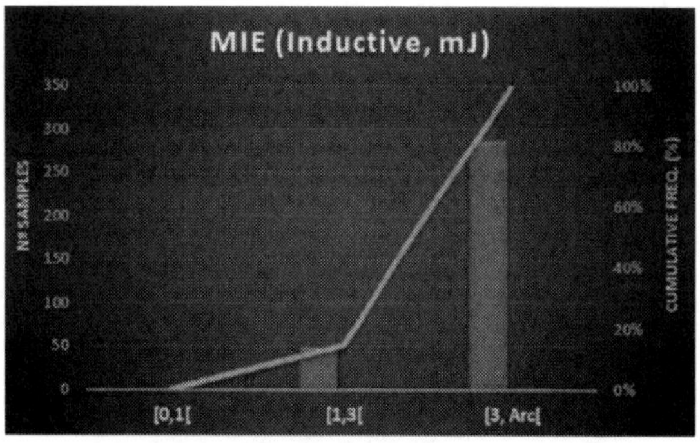

Organic dusts are highly sensitive to ignition

- MIE measures the minimum discharge of energy of an electrostatic or mechanical spark capable of igniting a dust cloud at r.t.
- The lower the MIE value, the higher the sensitivity of the material to ignition
- 335 organic dust MIE (inductive)
- **100% with MIE ≥1 mJ**

Minimum Ignition Temperature (MIT)

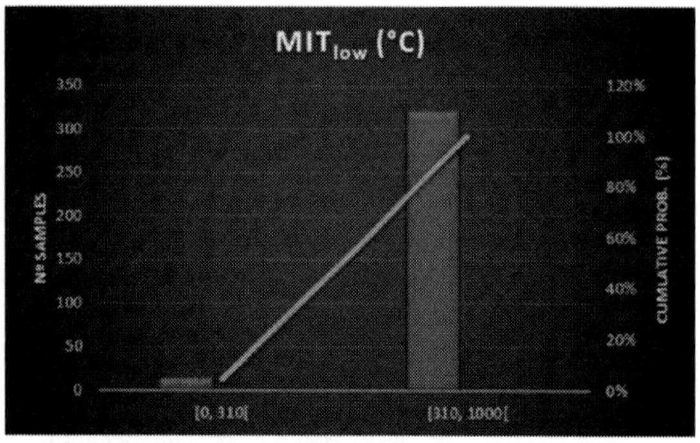

Organic dusts ignite at temperatures above 310°C

- MIT is the minimum temperature at which a dust cloud will auto-ignite.
- The MIT_{low} is the highest temperature at which the dust cloud does not ignite.
- 336 organic dusts MIT measurements
- **96% with MIT ≥310°C**

Combustibility and Burning Rate

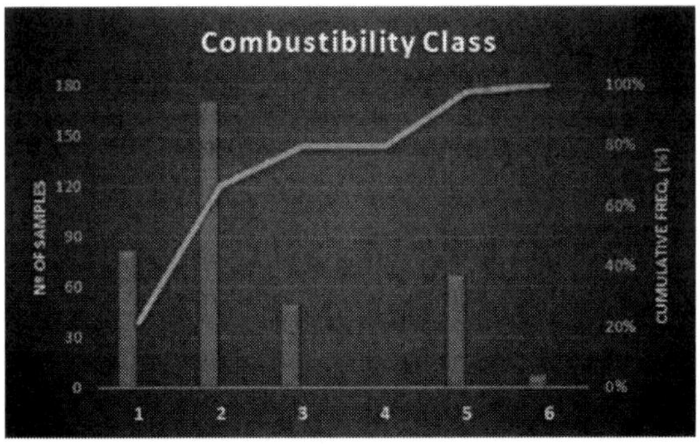

Organic dust are not classified as *Readly Combustible Solids*

- Combustibility and Burning Rate tests are designed to determine if, and to what extent, a fire started by an external source will propagate in a dust.
- 377 measurements
- **98% with combustibility class bellow 5**
- **None was classified as "Highly Flammable Solid"**

Resistivity Class

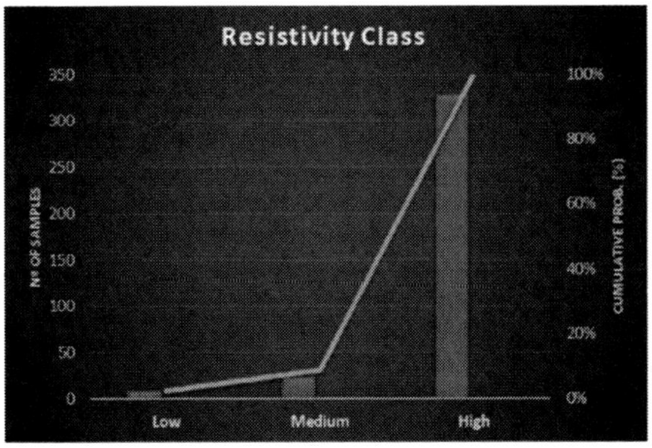

Organic dust are highly insulating

- Resistivity is a measure of the electrical resistance of a material and is the primary criterion for classifying dusts as Low
  - Low resistivity (<1MΩ.m)
  - Medium (1 MΩ.m – 10 GΩ.m)
  - Highly insulating (>10GΩ.m)
- Insulating materials retain electrostatic charge with potential to further discharges
- 359 resistivity measurements
- **91% classified as highly insulating**

Inherently Safe

Safer operations to remove the *probability* of accidents or minimize/reduce their consequences:

- ❑ **Reducing explosive atmosphere by LOC**
  - Inerted atmosphere in some cases with O_2 monitoring with SIL 1 or 2 instrumentation (based on LOPA assessments)
- ❑ **Preventing ignition sources**
  - Earthing and bonding equipment in some cases with Safety Instrumentation interlocking powder feed
  - Earthing of people with dissipative footwear
- ❑ **Preventing charge accumulation**
  - Equipments ATEX rated and made of conductive materials
  - Non-conductive joints properly bonded and grounded

Conclusions

- ❑ Efficient process safety information (PSI) management
- ❑ Working on the prevention of dust explosion. Equipments and operations inherently safe
- ❑ Sustainability improvement by reducing materials consumption
- ❑ Cost reduction
- ❑ Keeping up with Hovione safety standards

Safety First, Quality Always

Aknowlegments

This work resulted from the colaboration of all team member

Process Safety Scale-Up & Design Team

Bibliography

- Richter, S. M., *Process Safety Progress*, AIChE, 31, 2012, 165–169
- Amyotte, P., Eckhoff, R. *Journal of Chemical Health and Safety,* 17(1), 2010, 15–28
- Edwards, D. W., *Process Safety and Environmental Protection*, 83(B2), 2005, 90–100
- Zhao Y., Kingsly Ambrose R. P., *Powder Technology,* 408, 2022, 117704
- Kamarizan Kidam, K. *et al.*, Journal of Loss Prevention in the Process Industries, 2015, 1–12
- Paul R. Amyotte, Process Safety and Environment Protection, 87, 2009, 35–39

Any
Questions?

Elsa Gonçalves
edgoncalves@hovione.com

www.hovione.com

Hovione

In it for life

Thank you
for your attention

Elsa Gonçalves
edgoncalves@hovione.com

www.hovione.com

Hovione

In it for life

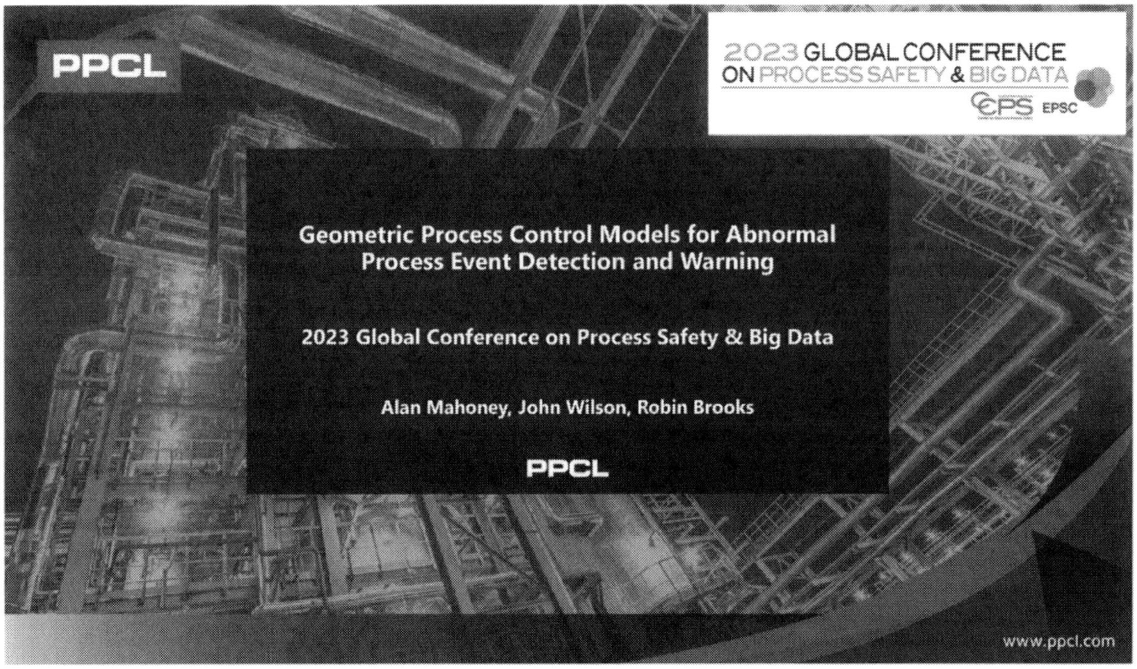
PPCL
2023 GLOBAL CONFERENCE ON PROCESS SAFETY & BIG DATA
CCPS EPSC
Geometric Process Control Models for Abnormal Process Event Detection and Warning
2023 Global Conference on Process Safety & Big Data
Alan Mahoney, John Wilson, Robin Brooks
PPCL
www.ppcl.com

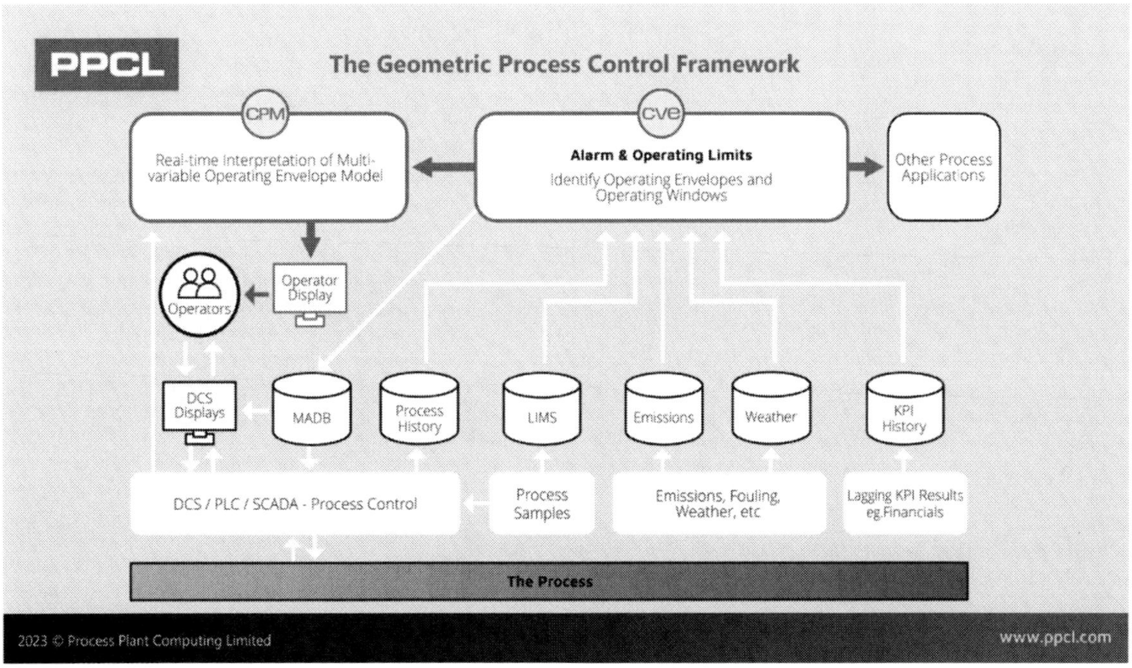
PPCL
The Geometric Process Control Framework
CPM
Real-time Interpretation of Multi-variable Operating Envelope Model
CVE
Alarm & Operating Limits
Identify Operating Envelopes and Operating Windows
Other Process Applications
Operators
Operator Display
DCS Displays
MADB
Process History
LIMS
Emissions
Weather
KPI History
DCS / PLC / SCADA - Process Control
Process Samples
Emissions, Fouling, Weather, etc
Lagging KPI Results eg.Financials
The Process
2023 © Process Plant Computing Limited
www.ppcl.com

PPCL

Abnormal Events

- Unplanned qualitative changes in operation
  - Faults
  - Disturbances
  - Accidents
- What do we want to do about it?
  - Recover
  - Mitigate
  - Repair
  - Understand
  - Improve
- What would help?
  - Early Detection
  - Direction
- Other technologies
  - Alarms
  - PCA/PLS
  - ML classifiers
  - Mechanistic models

PPCL

2-D Schematic of Operating Envelope

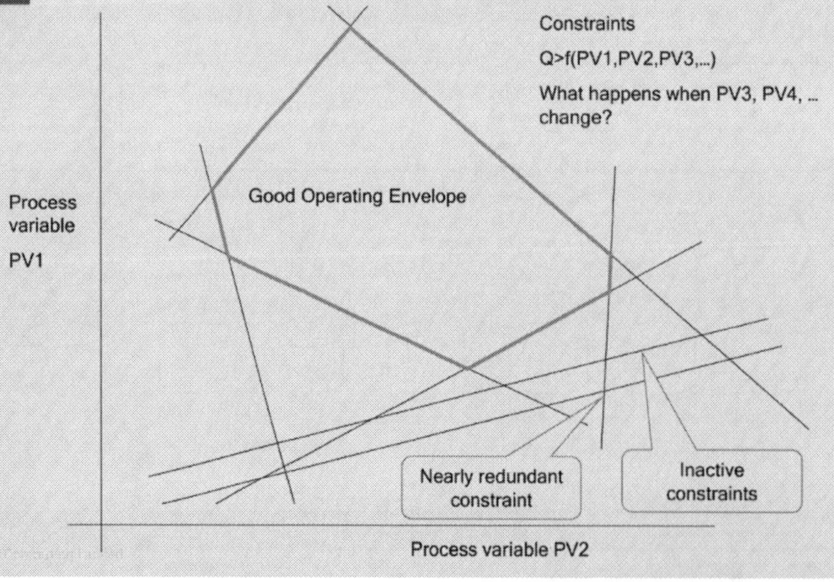

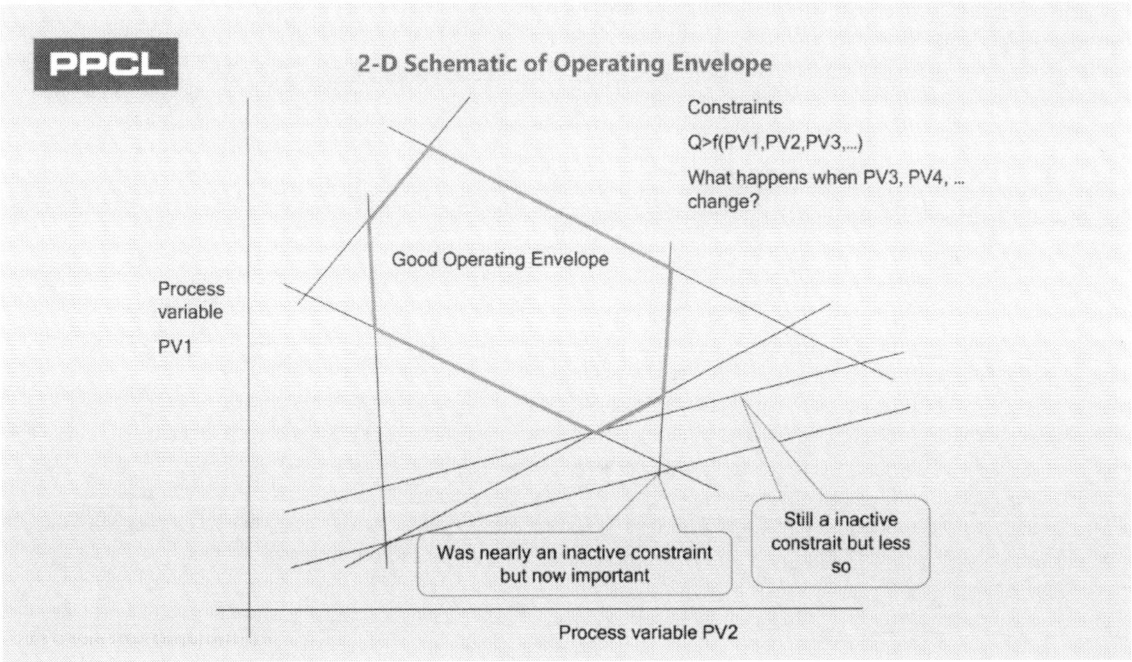
PPCL
2-D Schematic of Operating Envelope
Constraints
Q>f(PV1,PV2,PV3,...)
What happens when PV3, PV4, ... change?
Good Operating Envelope
Process variable PV1
Still a inactive constrait but less so
Was nearly an inactive constraint but now important
Process variable PV2

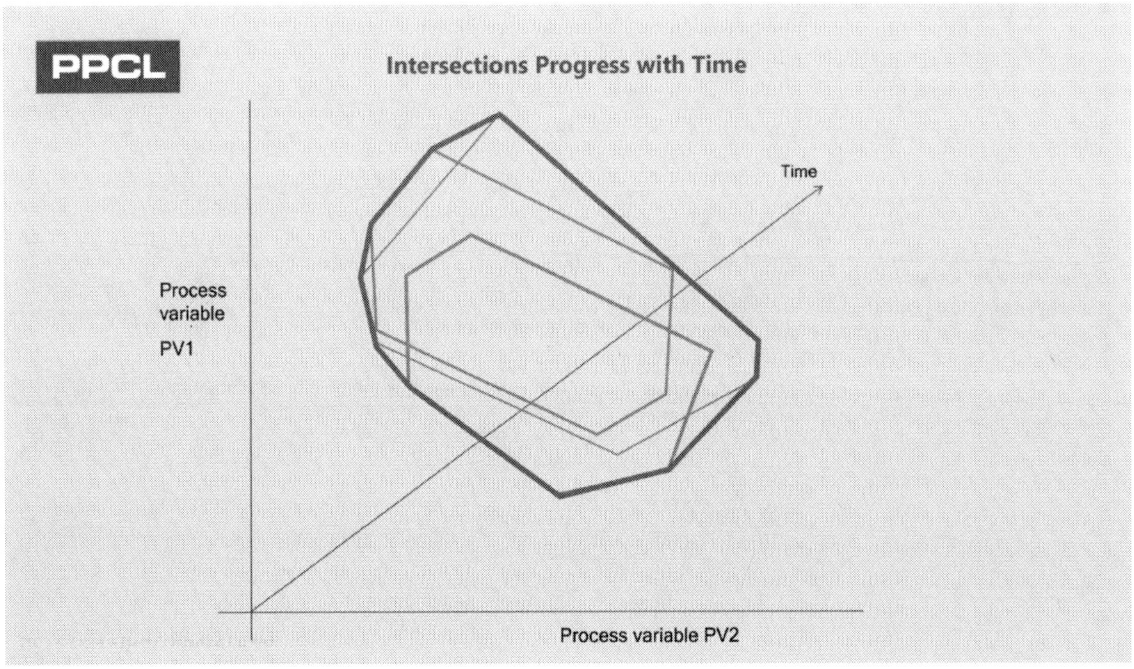
PPCL
Intersections Progress with Time
Time
Process variable PV1
Process variable PV2

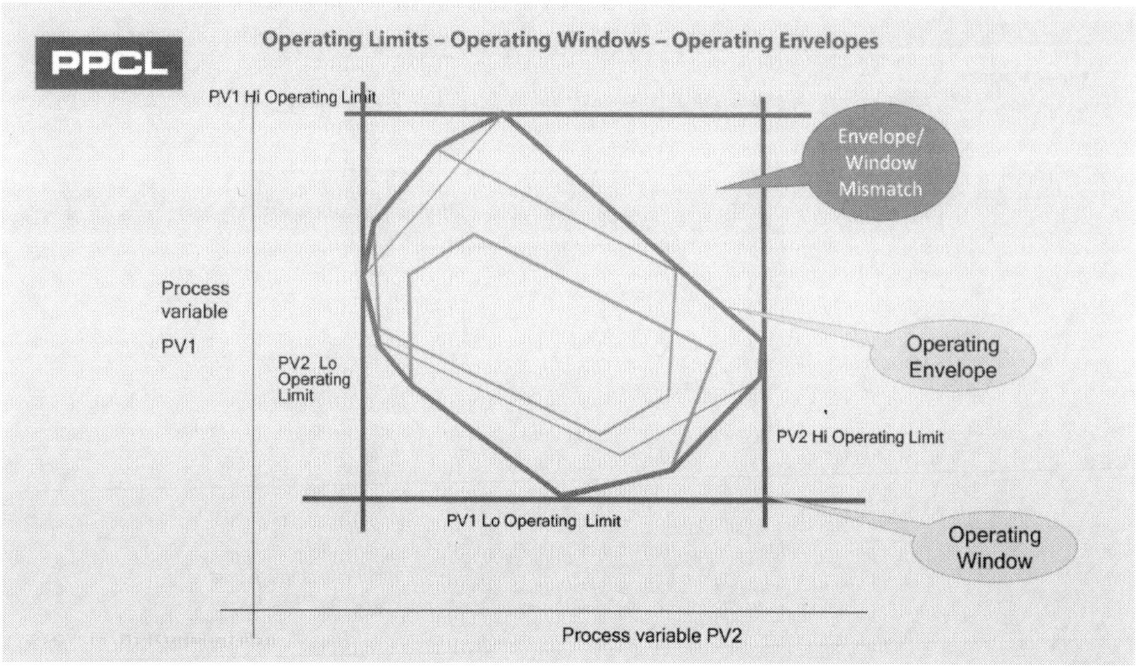
PPCL
Operating Limits – Operating Windows – Operating Envelopes
PV1 Hi Operating Limit
Envelope/ Window Mismatch
Process variable PV1
PV2 Lo Operating Limit
Operating Envelope
PV2 Hi Operating Limit
PV1 Lo Operating Limit
Operating Window
Process variable PV2

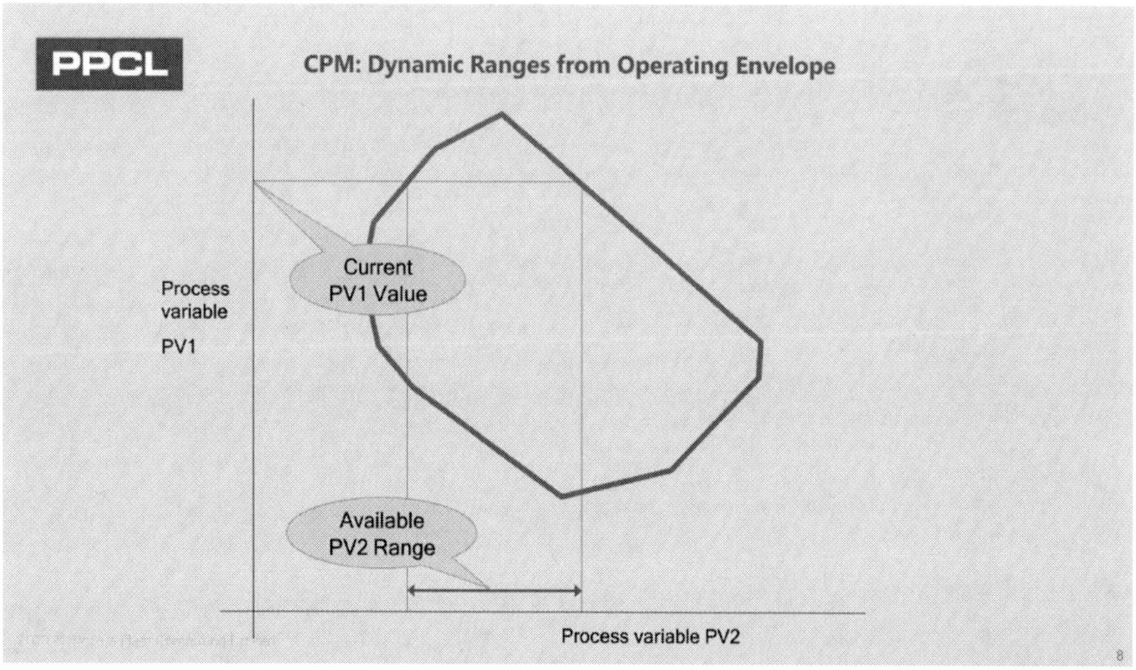
PPCL
CPM: Dynamic Ranges from Operating Envelope
Current PV1 Value
Process variable PV1
Available PV2 Range
Process variable PV2
8

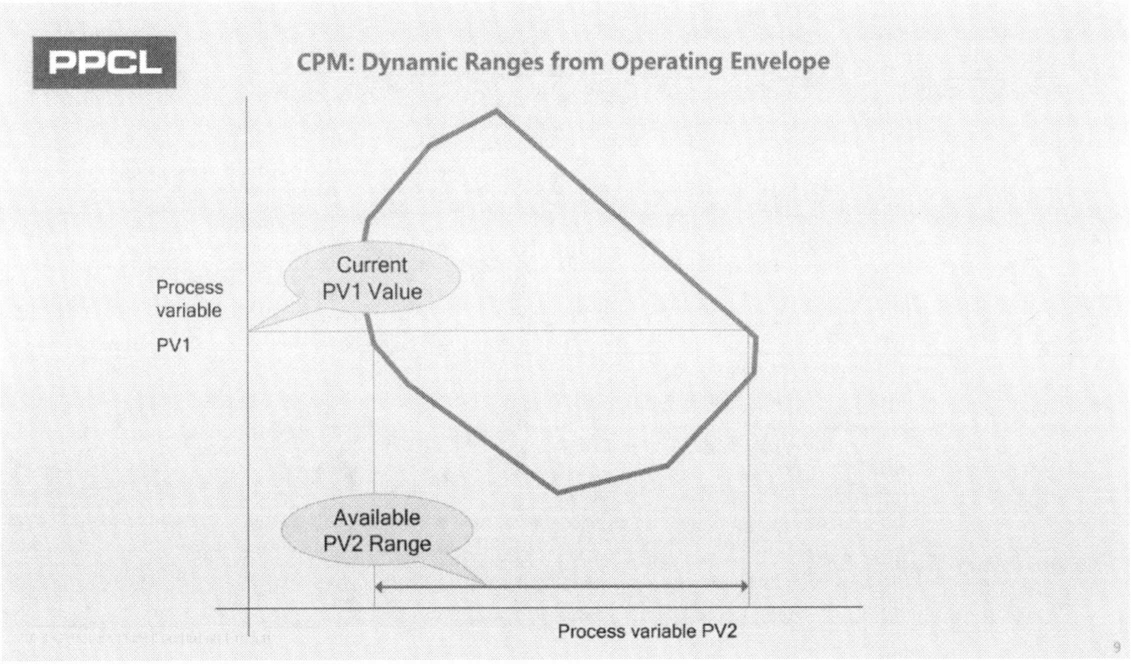
PPCL
CPM: Dynamic Ranges from Operating Envelope
Current PV1 Value
Process variable PV1
Available PV2 Range
Process variable PV2
9

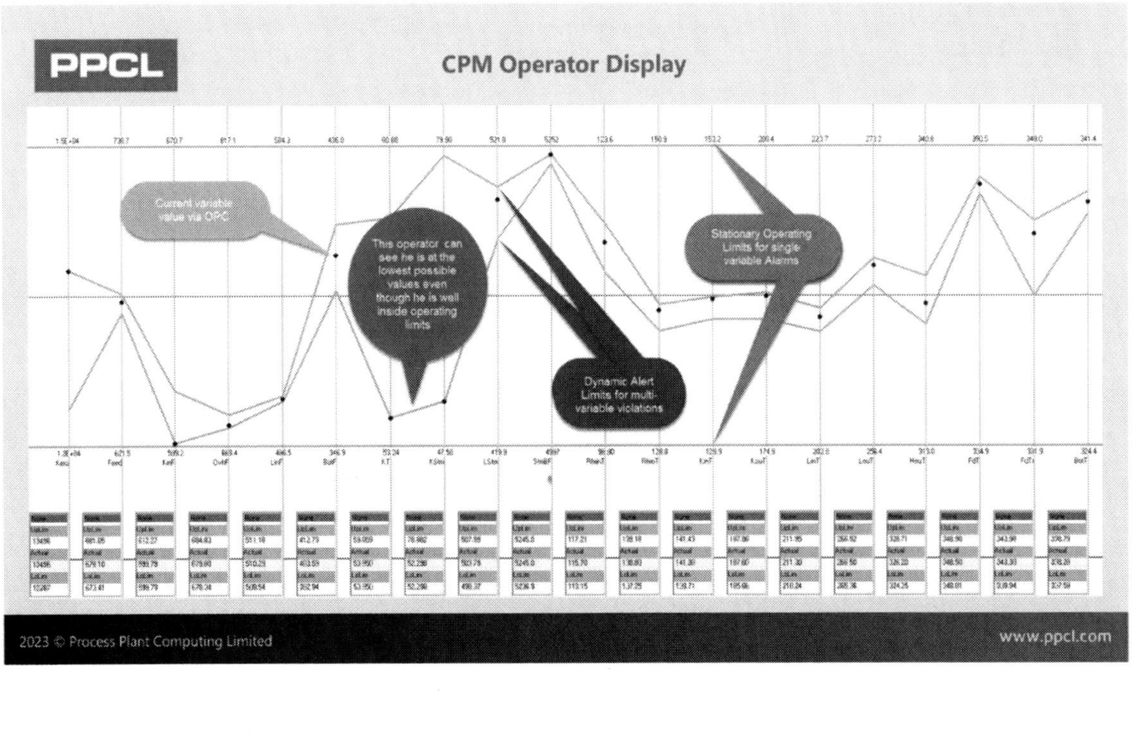
PPCL
CPM Operator Display
Current variable value via OPC
This operator can see he is at the lowest possible values even though he is well inside operating limits
Dynamic Alert Limits for multi-variable violations
Stationary Operating Limits for single variable Alarms
2023 © Process Plant Computing Limited
www.ppcl.com

PPCL

Demo System: Turbine and Generator

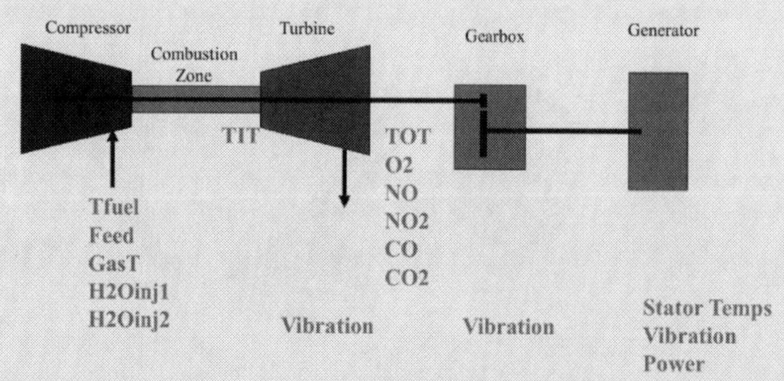

www.ppcl.com

Envelope of Normal Operation

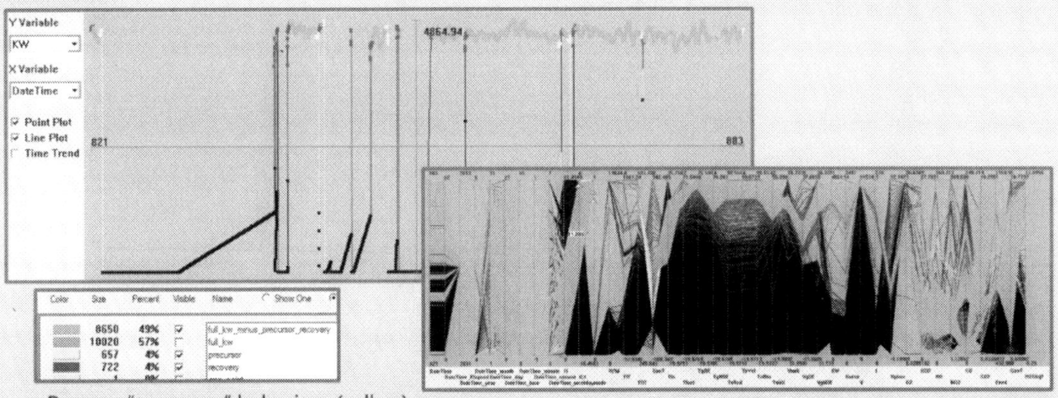

- Remove "precursor" behaviour (yellow)
- Remove "startup" transient behaviour (blue)
- Remove offline periods (black)
- Remaining (orange) is a normal operating envelope

 www.ppcl.com

Single Variable Event Detectors

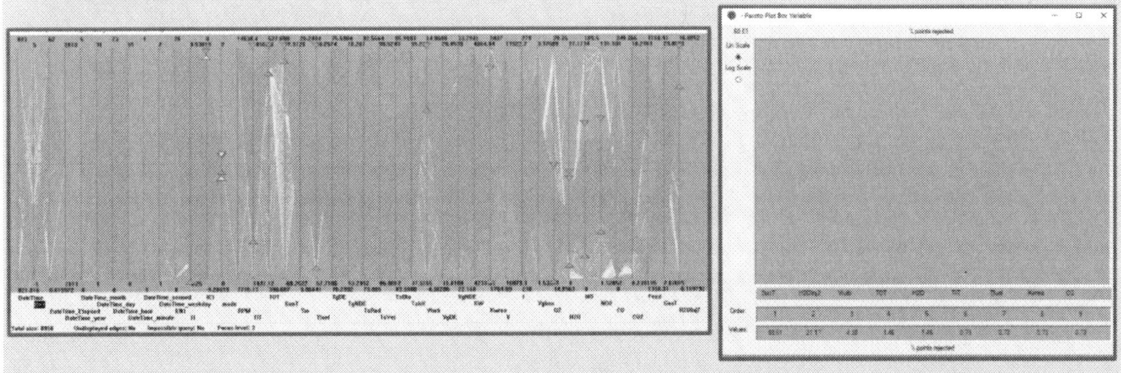

- Where normal operation differs most from historic event precurors
- Key variables to watch and box/envelope/alarm limits (cyan triangles)
- Two events detected >10 minutes out. Can we do better?

 www.ppcl.com

Operating Envelope Could Do Better

- If power Feed and GasT are imbalanced, yellow at bottom left/top right, could indicate precursor to a trip!
- Can't capture this with ranges on single variables.

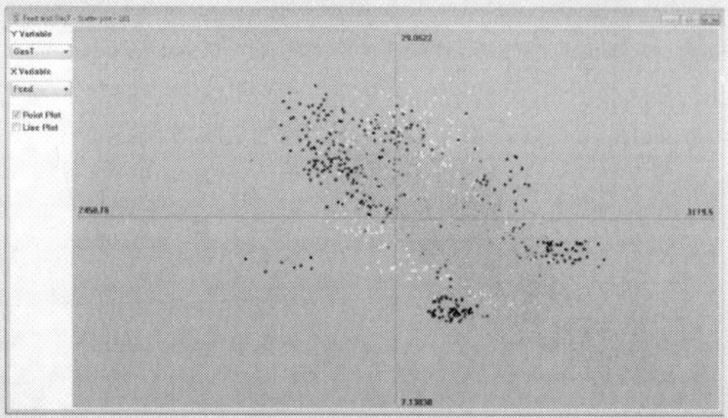

Building a GPC model

1. Using Visual Explorer, decide with hindsight where the plant should have operated by applying objectives (e.g. normal operation) as a query.

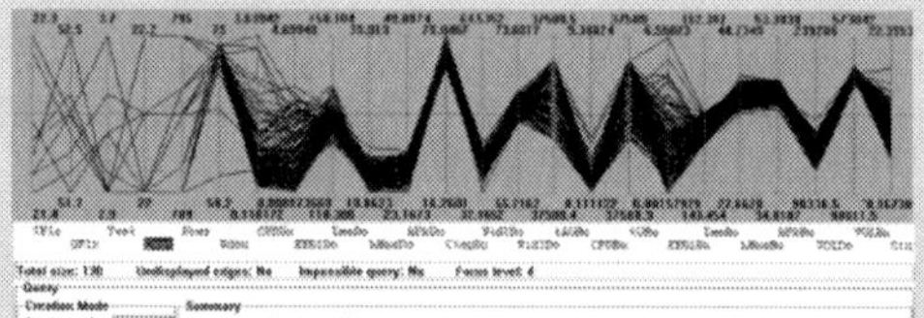

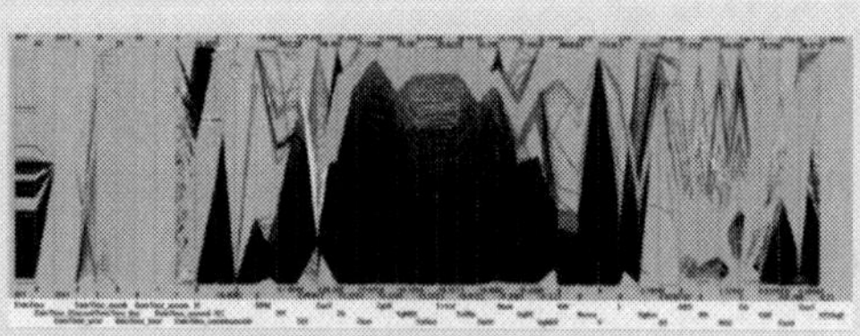

2. 'Focus' on the selected points, remove unwanted variables and save the points. These represent best past experience to be targeted in the future.

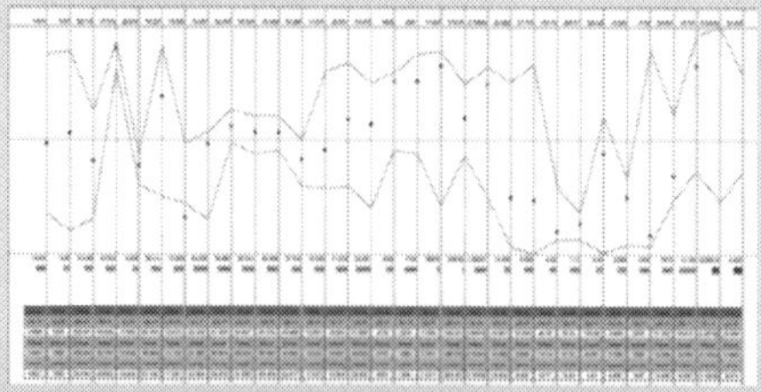

3. Open the file in Process Modeller. In real time, the current operation and envelope limits are shown to the operator

PPCL

Operators view – Gas Turbine/Generator Condition Monitoring

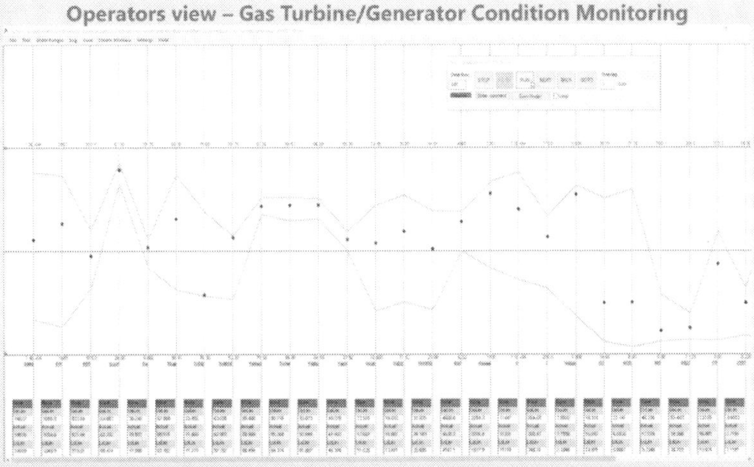

- Operators' display compressed: six hours in 72 seconds
- Warning 5 hours before upcoming event; individual variables give only 5-10 minutes

 www.ppcl.com

Ethylene Process

- Operation near surge offers best efficiency
- Process and external disturbances have potential to push compressor toward surge
- JT valves near compressor protect from surge but process can take hours to recover following activation
- Operators manage process demand and false load to balance these

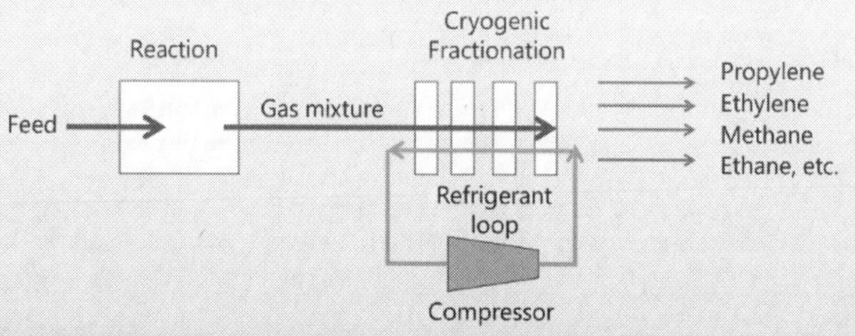

Building the Geometric Model

- Create an envelope from past operation far from fault conditions (yellow)
- Avoid entering the fault condition black space
- Violations of the envelope in real-time increase in the turquoise space when approaching the fault condition
- Operations model, need warning times of many minutes so operator can act

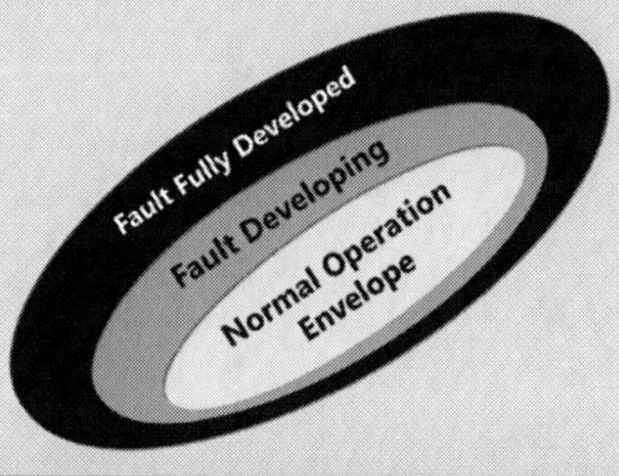

Model and Test Data

- Model lies within precursor data
- Single variable excursions provide little warning time

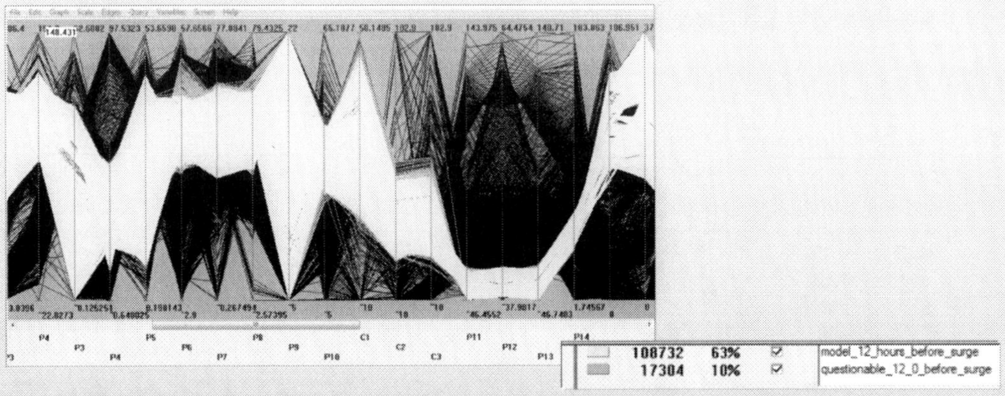

Early Surge Warning

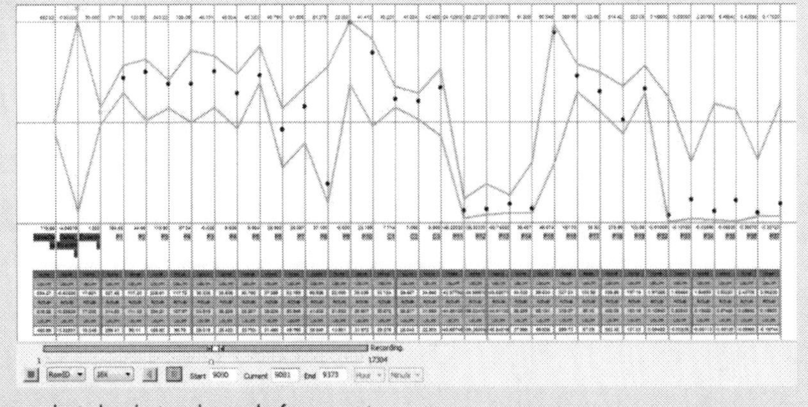

- Operator alerts begin two hours before event
- Alerts point to the variables that are key in understanding the event
- Opportunity for operator to avoid or mitigate the process impacts

PPCL

Compressor Event Performance

- Model run on 30 historic events
- 40% of surge events seen at least 15 minutes in advance
- 80% of surge events seen over 5 minutes in advance
- Extra time for operators to understand situation and make corrections

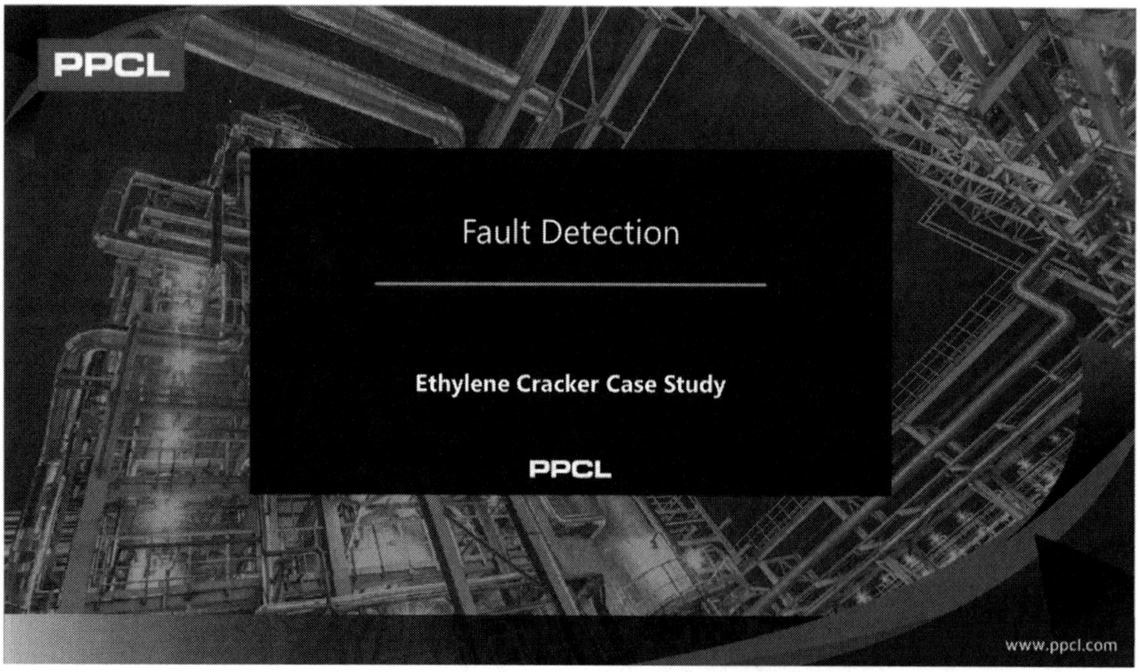

Ethylene Cracker Digital Twin

- 2 days 12 hours before event: Normal Operation

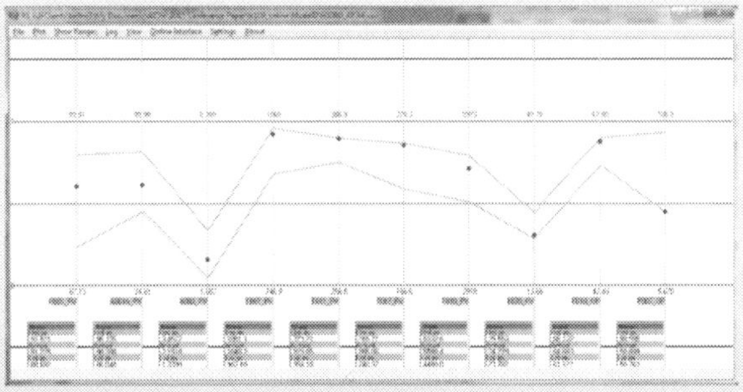

Reproduced courtesy of the author from "A Novel Approach to the Event Prediction and Mitigation Problem in an ethylene plant", Bell, M., Paper 098D, Proceedings of the AIChE Spring Conference 2015

www.ppcl.com

Ethylene Cracker Digital Twin

- 2 days 11 hours before event: first deviation detected

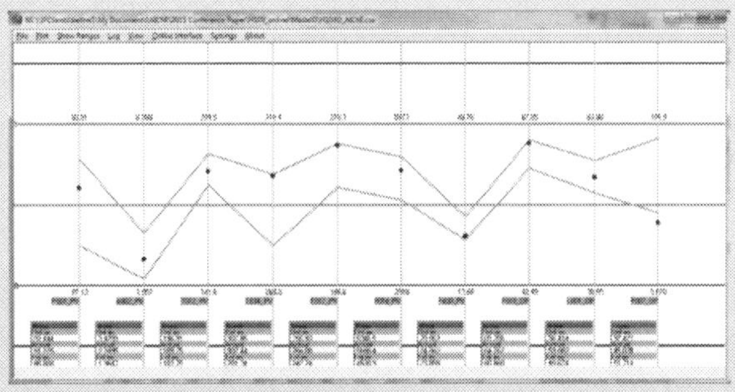

Reproduced courtesy of the author from "A Novel Approach to the Event Prediction and Mitigation Problem in an ethylene plant", Bell, M., Paper 098D, Proceedings of the AIChE Spring Conference 2015

www.ppcl.com

PPCL

Ethylene Cracker Digital Twin

- 1 day before event: large deviation

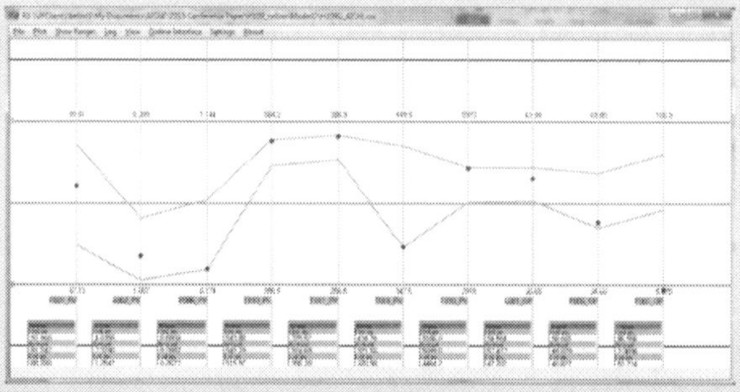

Reproduced courtesy of the author from "A Novel Approach to the Event Prediction and Mitigation Problem in an ethylene plant", Bell, M., Paper 098D, Proceedings of the AIChE Spring Conference 2015

PPCL

Ethylene Cracker Digital Twin

- 2 minutes before event: large deviations on three tags

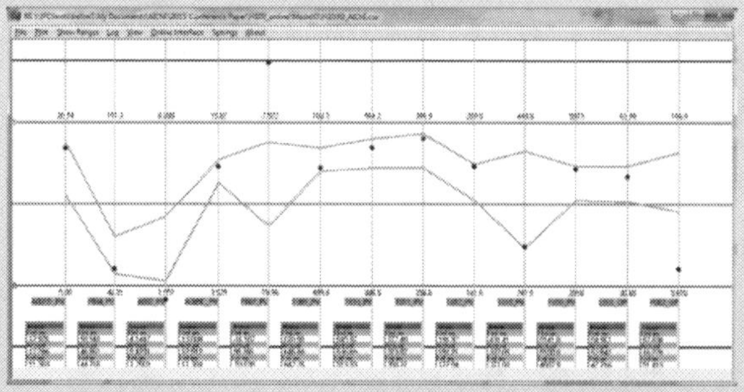

Reproduced courtesy of the author from "A Novel Approach to the Event Prediction and Mitigation Problem in an ethylene plant", Bell, M., Paper 098D, Proceedings of the AIChE Spring Conference 2015

PPCL

Ethylene Cracker Digital Twin

- During event: multiple alerts as operator brings furnace down

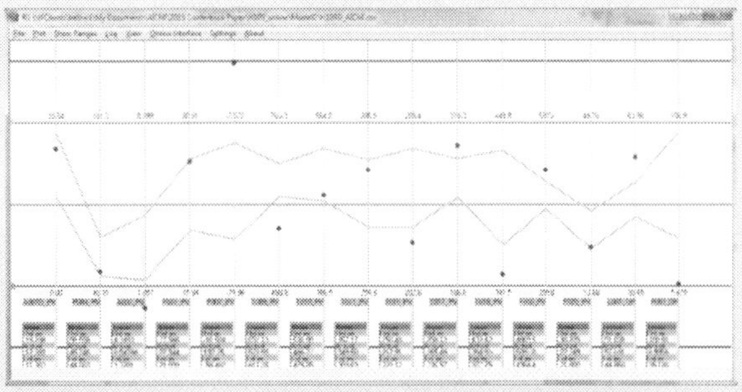

Reproduced courtesy of the author from "A Novel Approach to the Event Prediction and Mitigation Problem in an ethylene plant", Bell, M., Paper 098D, Proceedings of the AIChE Spring Conference 2015

 www.ppcl.com

PPCL

Benefits of Early Event Detection

"The method successfully identified a historical significant furnace event almost two and one half days before the event occurred, flagging the exact instrument that was failing. The application has identified 3 significant events during its first 8 months of operation.

"CPM has been actively notifying the console operator of any abnormal process conditions to the furnace with an on stream time of 99.9%."

 www.ppcl.com

PPCL

Benefits of Geometric Process Control Models

- Acheivable
  - No specialist math skills needed;
  - Models built & maintained by engineers close to process
- Extends condition monitoring to smaller equipment
  - Trip of a pump causes process disturbance that can take hours of costly settling out
  - Data to build models already present in historian
- Early indication of developing faults gives operators time to think and act
  - Better performance than PCA and PLS models
  - Indication of key variables in developing situation

PPCL

The Geometric Process Control Framework

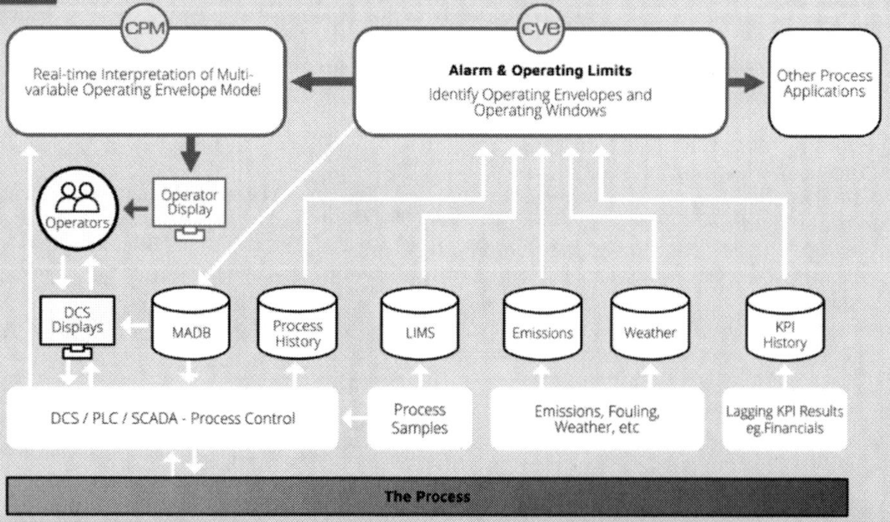

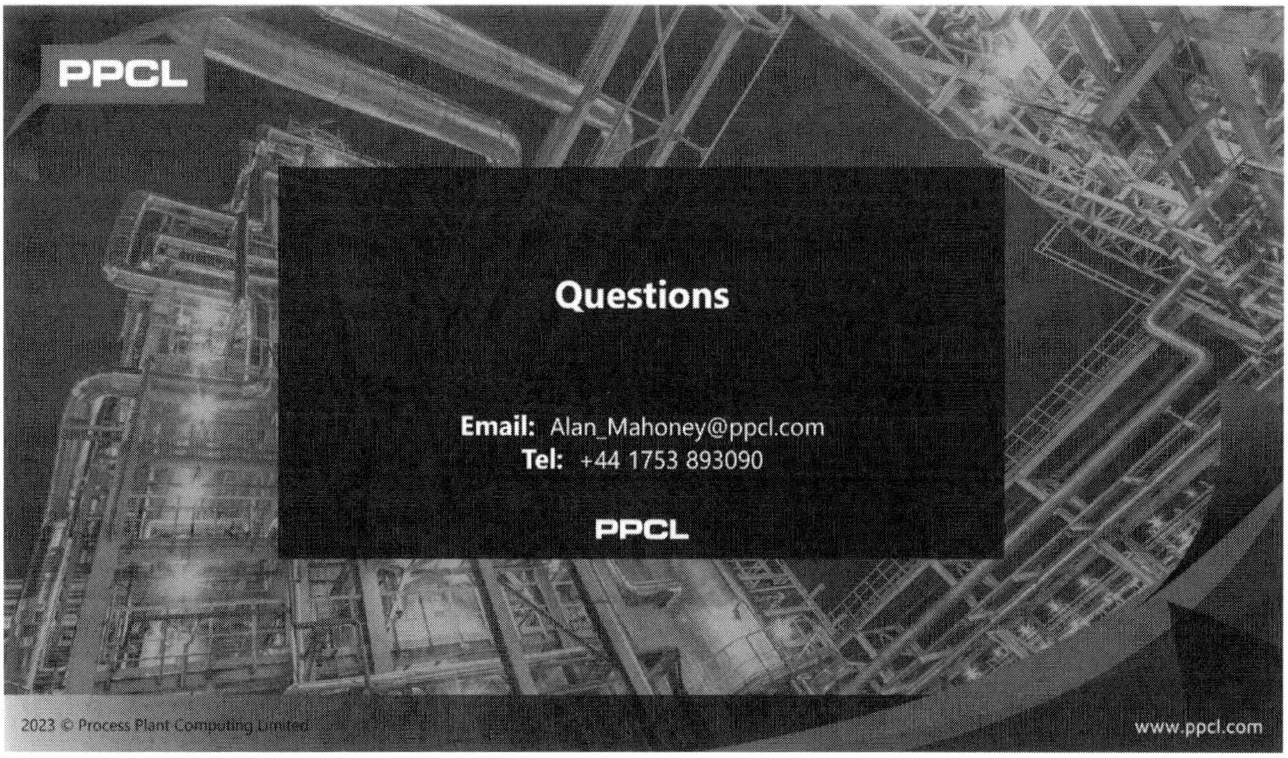
PPCL
Questions
Email: Alan_Mahoney@ppcl.com
Tel: +44 1753 893090
PPCL
2023 © Process Plant Computing Limited
www.ppcl.com

Harnessing the Power of Digitalization

to Better Manage Process Safety Health

October 25, 2023

2023 GLOBAL CONFERENCE ON PROCESS SAFETY & BIG DATA

CCPS EPSC

Process Safety Health

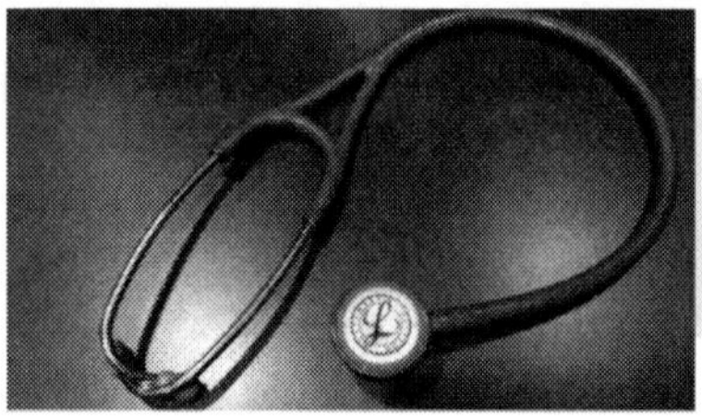

***Yesterday's** readings*

Medical practitioners check patients' vital signs as a proactive method to monitor their health risks. In the field of medical science, technology advances rapidly, as there is little resistance to change. Today dynamic readings are the norm.

***Today's** live readings*

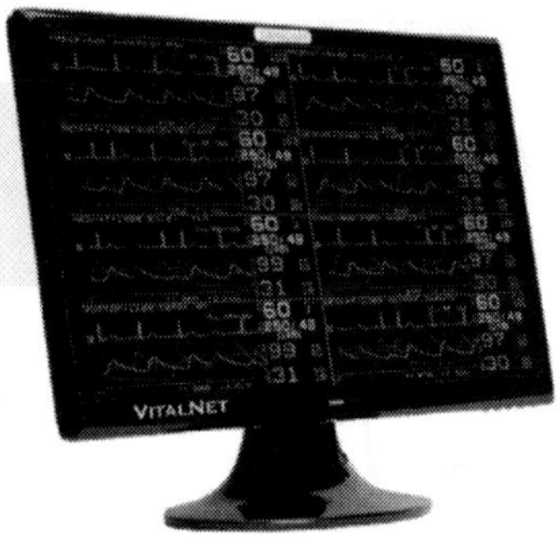

Similarly, process safety management of production facilities requires accurate barrier health monitoring. However, as of 2023, only 13% of organizations report making full use of technology solutions available for compliance & improved predictive decision-making.

Despite better technology and more resources...

Roughly 65%

Still see a gap between safety goals and realities.

Only 13%

Take full advantage of the range of tools for compliance and predictive decision-making.

8th ANNUAL Safety Survey Report

sphera 2

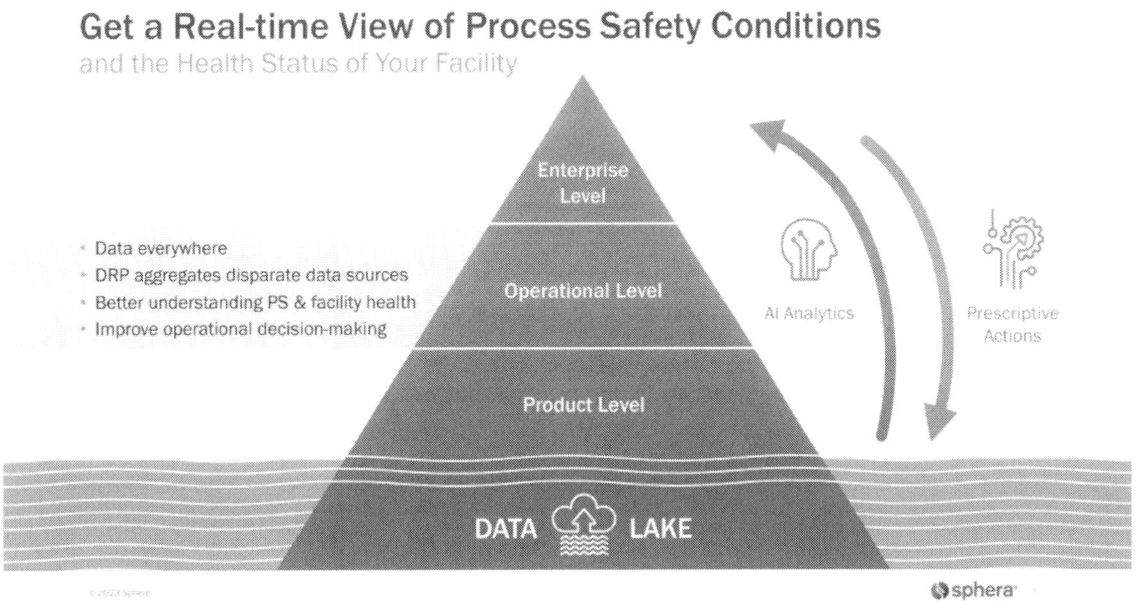
Get a Real-time View of Process Safety Conditions
and the Health Status of Your Facility
Data everywhere
DRP aggregates disparate data sources
Better understanding PS & facility health
Improve operational decision-making
Enterprise Level
Operational Level
Product Level
DATA LAKE
AI Analytics
Prescriptive Actions
sphera

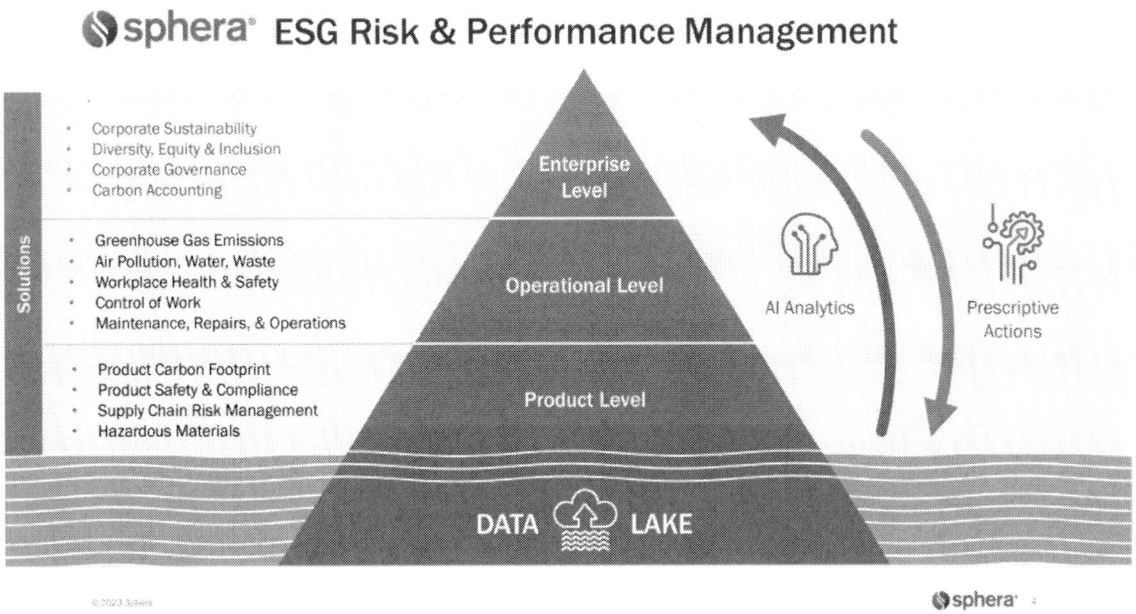
sphera ESG Risk & Performance Management
Solutions
Corporate Sustainability
Diversity, Equity & Inclusion
Corporate Governance
Carbon Accounting
Greenhouse Gas Emissions
Air Pollution, Water, Waste
Workplace Health & Safety
Control of Work
Maintenance, Repairs, & Operations
Product Carbon Footprint
Product Safety & Compliance
Supply Chain Risk Management
Hazardous Materials
Enterprise Level
Operational Level
Product Level
DATA LAKE
AI Analytics
Prescriptive Actions
sphera

Understand What's Driving the Risk

In real-time, on the next shift, tomorrow, and beyond.

Understand the hour-by-hour impact of cumulative risk by major hazard event.

See the real-time health of the impacted barriers.

Navigate from shift-to-shift to see the evolution of process safety barrier impact & cumulative risk.

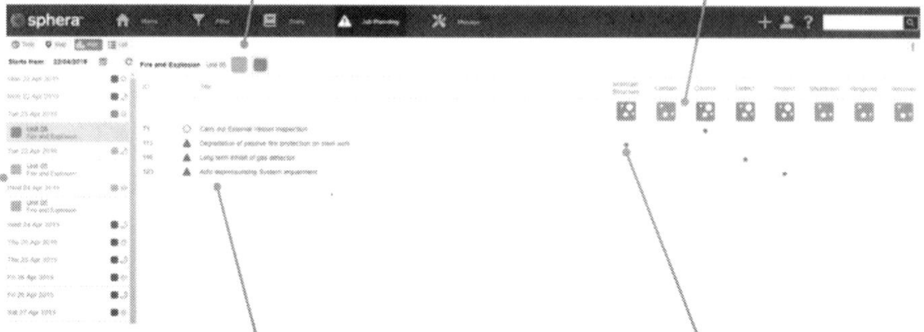

Click to reveal the details of equipment health and deviations.

See how equipment health and deviations impact process safety barriers.

sphera

Dynamic Bow Tie = Process Safety Health

Traditional bow tie is **static**

Lives in a file.

DRP bow tie:

- Reads DATA within a system
- Gives an early warning of impairment (potential near miss)
- Changes dynamic of maintenance prioritization
- OMT clear view of the current situation

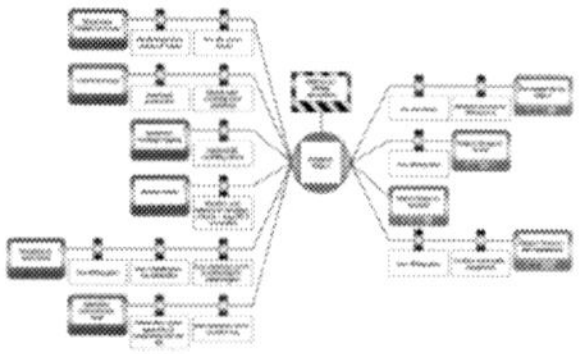

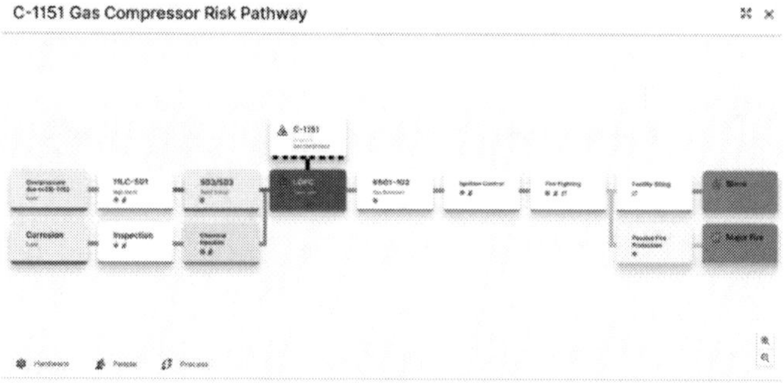

sphera

Benefits

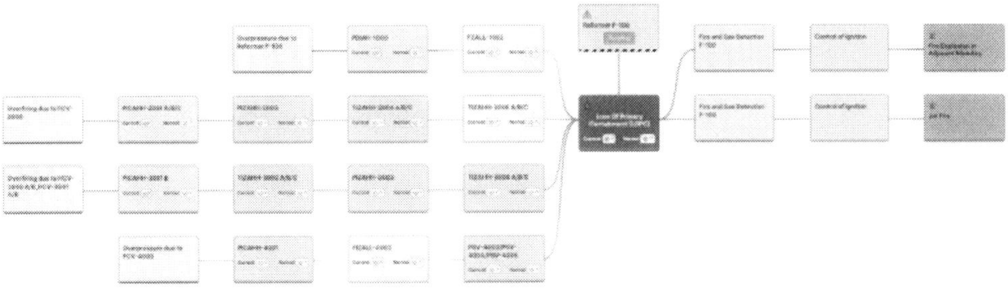

- Provides a clear view of the current situation
- Identify the Probability of a Top Event happening
- Gives an early warning of Impairment (potential near miss alerts)
- Changes the dynamic of Maintenance prioritization

sphera

Cumulative Risk UI

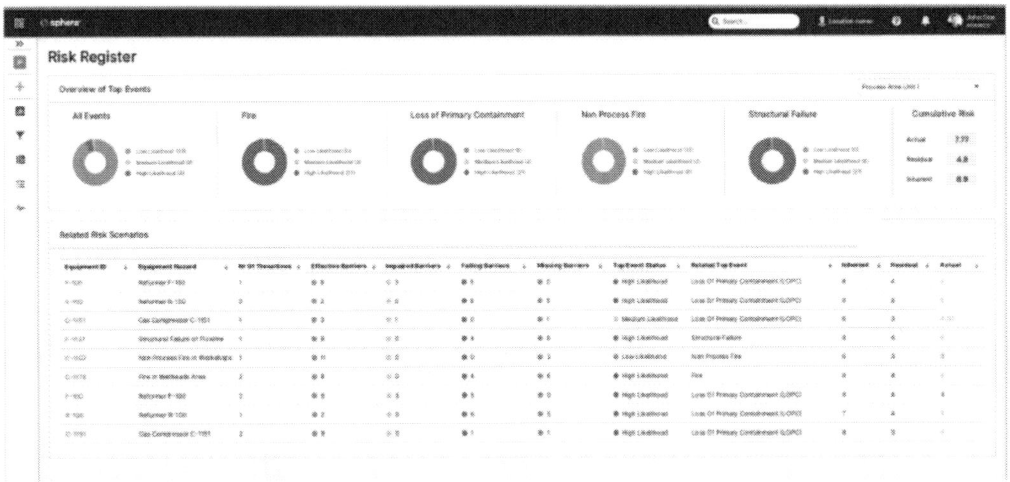

sphera

"I Know my plant is safe"

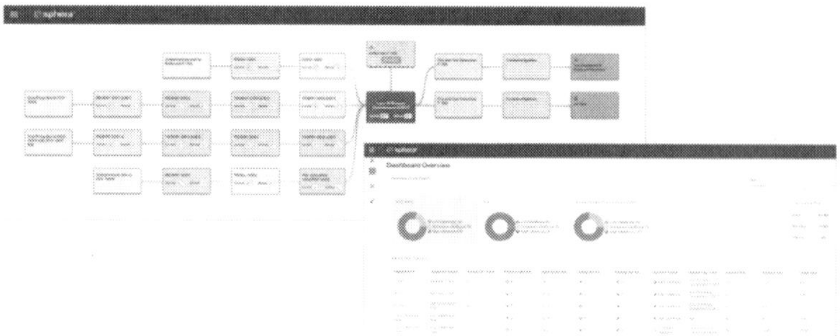

Initial Risk

↓

Residual Risk

↓

Actual Risk

Data at fingertips

Connected through IIOT

Clear Linear visualization

Continuous Improvement loop

Improve Resource Utilization and Increase Uptime

Increased visibility helps you improve maintenance prioritization, asset integrity, safety assurance and production uptime.

With DRP, you can reduce risk and disruption and more easily realize the value of your projects.

Dynamic Risk Pathways enables improved risk management, enhanced workforce engagement in EHS and better utilization of contractors.

Conclusion

Dynamic Risk Pathways monitoring influences established decision-making processes such as maintenance prioritization by taking **risk** into consideration as a critical contributor. The solution enables companies to:

Show the near real-time status of each barrier and how it interacts with other barriers.	*Calculate the probability of a top event happening.*	*View the status of the mitigation and recovery barriers.*	*Understand the cumulative risk of specific scenarios.*	*Get an aggregated area risk value for a specific level of the site hierarchy.*	*Make information accessible and understandable throughout the organization, from the workers to executives.*

Top event Incidents cause **disruption, lost production** and **repair costs.**

$$$ **Prevention Technology** is more cost effective $$$

CCPS Evolution of Risk Based Process Safety

Standards-based strategy	Compliance-based strategy	Continuous improvement-based strategy	Risk-based strategy	Visible & Dynamic Risk
What Should I do?	What do I Have to do?	How can I use my experience to Improve?	How can I Better Manage Risk?	How can I make full use of Technology
Prescriptive	*Minimum requirements*	*Use of lagging indicators*	*Risk information + leading indicators*	*My plant is safe and I know it*

Future → Visible & Dynamic Risk

→ Here now!

Please Feel Free to Contact Me on LinkedIn or Email

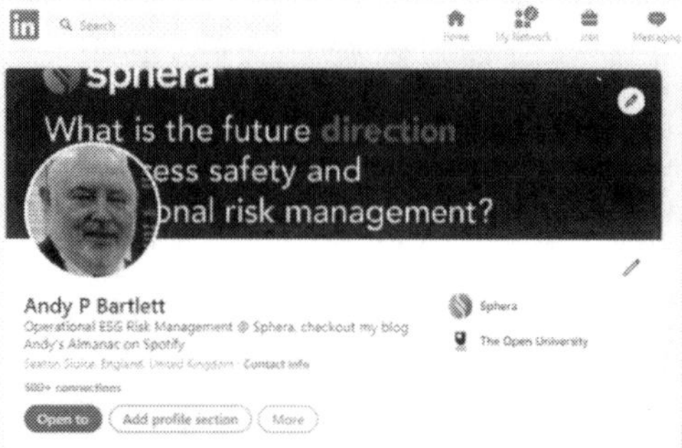

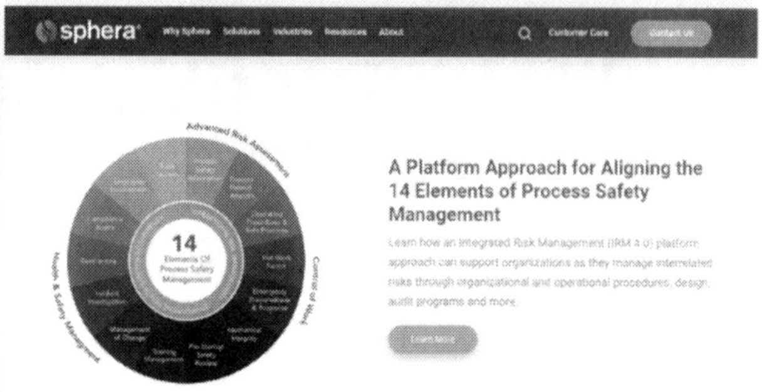

Email: Abartlett@sphera.com

Thanks for listening!

Address 130 East Randolph St, Suite 1900 Chicago, IL 60601 USA

Phone (312) 796-7160

Online sphera.com/contact-us

sphera Why Sphera Solutions Industries Resources About Customer Care Contact Us

By Sphera's Editorial Team | October 26, 2020

Barriers and safeguards are in place to help hazardous industries prevent incidents, but being able to monitor all those barriers at one time across an organization has always been difficult, especially how different key controls interact—until now. Dynamic Risk Pathways is the next-generation Process Safety Management software powered by operational, process and equipment data that drives home results for safe operations in real time.

2023 GLOBAL CONFERENCE
ON PROCESS SAFETY & BIG DATA

CCPS EPSC

2023 GLOBAL CONFERENCE
ON PROCESS SAFETY & BIG DATA

CCPS EPSC

How Enhanced Reality technology reduces human errors through increased operator competency

Dr. Susanna Voges
Voovio Technologies SL

voovio

Co-organized by CCPS & EPSC

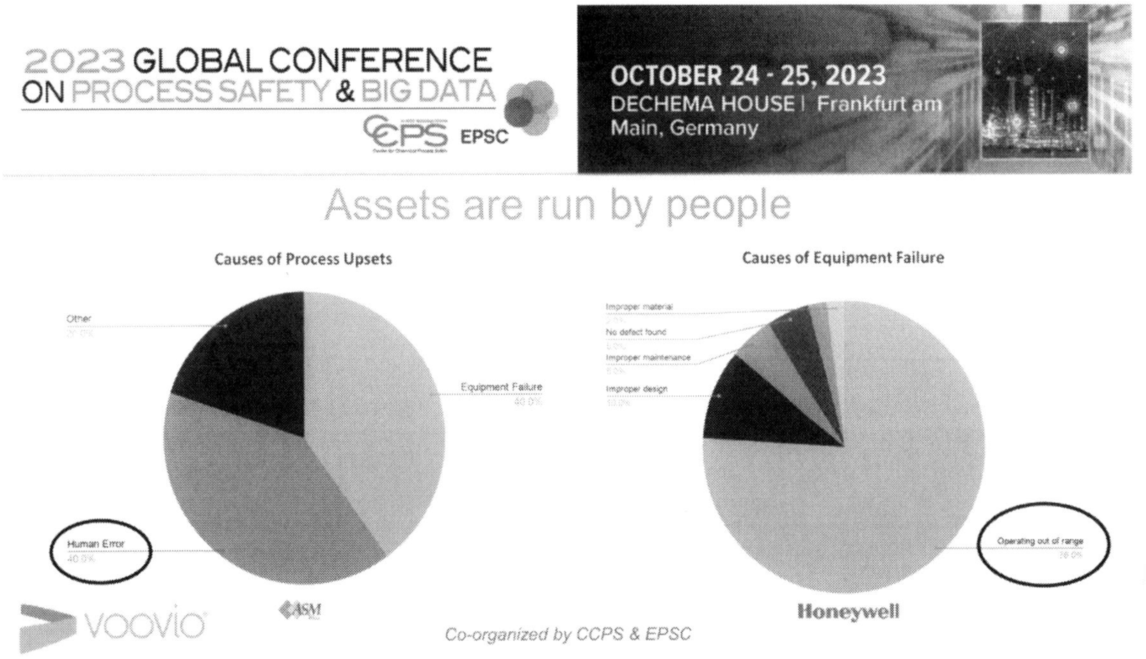
2023 GLOBAL CONFERENCE
ON PROCESS SAFETY & BIG DATA
CCPS EPSC
OCTOBER 24 - 25, 2023
DECHEMA HOUSE | Frankfurt am Main, Germany
Assets are run by people
Causes of Process Upsets
Other
Equipment Failure
40.0%
Human Error
40.0%
Causes of Equipment Failure
Improper material
No defect found
Improper maintenance
Improper design
Operating out of range
Honeywell
VOOVIO
Co-organized by CCPS & EPSC

2023 GLOBAL CONFERENCE
ON PROCESS SAFETY & BIG DATA
CCPS EPSC
OCTOBER 24 - 25, 2023
DECHEMA HOUSE | Frankfurt am Main, Germany
Digital twins are not complete until they capture the influence of the person in the operation
Discrete Digital Twin
Asset
Person
Gartner
Discrete Digital Twin of the Person*
*Gartner: Emerging Technologies: Revenue Opportunity Projection of Digital Twins
VOOVIO
Co-organized by CCPS & EPSC

2023 GLOBAL CONFERENCE ON PROCESS SAFETY & BIG DATA

CCPS EPSC

OCTOBER 24 - 25, 2023
DECHEMA HOUSE | Frankfurt am Main, Germany

Let's see a demo

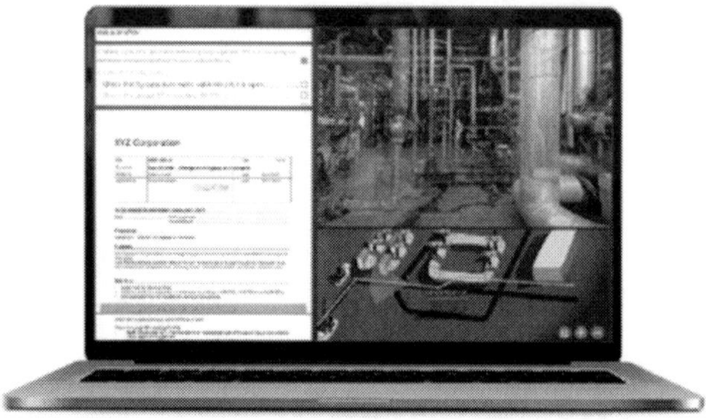

VOOVIO

Co-organized by CCPS & EPSC

2023 GLOBAL CONFERENCE ON PROCESS SAFETY & BIG DATA

CCPS EPSC

OCTOBER 24 - 25, 2023
DECHEMA HOUSE | Frankfurt am Main, Germany

Procedure Clarity & Competency

VOOVIO

Co-organized by CCPS & EPSC

OCTOBER 24 - 25, 2023
DECHEMA HOUSE | Frankfurt am Main, Germany

What is Procedure Clarity for an Operator?

1. Clear and concise actions
2. Ability to visualize procedure steps in the field
3. Ability to practice standardized execution (*learn by doing*)

Procedure Analysis.

We randomly sampled and analyzed ~300 customer procedures:

S/U and S/D incl Equipment Swaps	83%
Normal	8%
Emergency/Abnormal	7%
Maintenance Overhaul	1%
Temporary	1%

90% Of SOPs analyzed contained ambiguities

Two reasons for suboptimal procedure clarity.

1. Procedure does not reflect operator practice

Written procedures vary from the unit's established practice with some practices varying from shift to shift either due to training inconsistencies or a lack of standardized execution.

2. Ambiguities and generalizations

Procedures are written with a lot of technical content, generalizations and ambiguities that assume clear understanding by the operator.

2023 GLOBAL CONFERENCE ON PROCESS SAFETY & BIG DATA

CCPS EPSC

OCTOBER 24 - 25, 2023
DECHEMA HOUSE | Frankfurt am Main, Germany

1 step

2.3 *Open the non-instrumented fuel gas cock valves and air dampers.*

→

10 actions

- Interacting with 5 air dampers
- Actuating 5 fuel gas valves

A very common example.

VOOVIO

Co-organized by CCPS & EPSC

2023 GLOBAL CONFERENCE ON PROCESS SAFETY & BIG DATA

CCPS EPSC

OCTOBER 24 - 25, 2023
DECHEMA HOUSE | Frankfurt am Main, Germany

Also common and very open-ended...

1 step → 53 actions

Actuating 15 different valves

MANUALLY FLUSH all CTO's, top/bottom flushes, bypasses around flush flow meters, and any other isolation flush valves

CAUTION
Failure to properly flush all CTOs may result in a reactor poisoning event

VOOVIO

Co-organized by CCPS & EPSC

'Line walk' and simulator improve procedure clarity

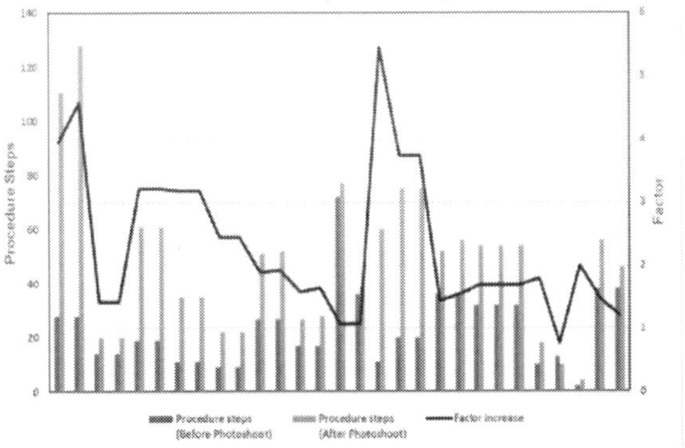

Voovio Line Walk process eliminates procedure generalities and ambiguities resulting in clear and concise action steps.

On average, procedure steps increase by a factor of 2.4

Simulated SOP can be reviewed by teams to further improve SOP accuracy during table-top reviews or post incidents.

voovio

2023 GLOBAL CONFERENCE ON PROCESS SAFETY & BIG DATA
CCPS EPSC
OCTOBER 24 - 25, 2023
DECHEMA HOUSE | Frankfurt am Main, Germany

Procedure Clarity Results in Competency

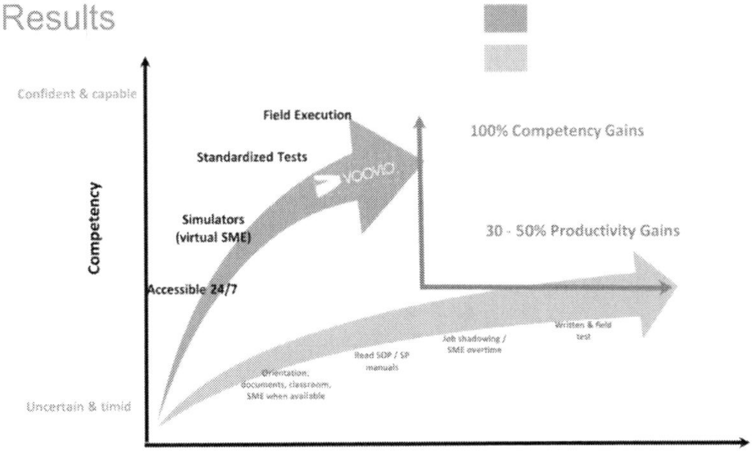

voovio

Operator Feedback

In surveys conducted with end users (operators) between 2018-2020, all respondents ***"Strongly Agreed"*** Procedure Simulators,

- make it easier to learn and understand procedures than their current (traditional) training methods
- provide clarity on aspects of their job that they did not have before
- will increase compliance with operating procedures in their area
  - Procedures are accurate and action-based
  - Visual clarity
  - Procedures and practices are standardized

VOOVIO

Co-organized by CCPS & EPSC

2023 GLOBAL CONFERENCE ON PROCESS SAFETY & BIG DATA
CCPS EPSC
OCTOBER 24 - 25, 2023
DECHEMA HOUSE | Frankfurt am Main, Germany

Workforce Performance Data

- Data from 14 customers
- 1 critical SOP was assigned to each **qualified** operator
- 1 month of autonomous practice

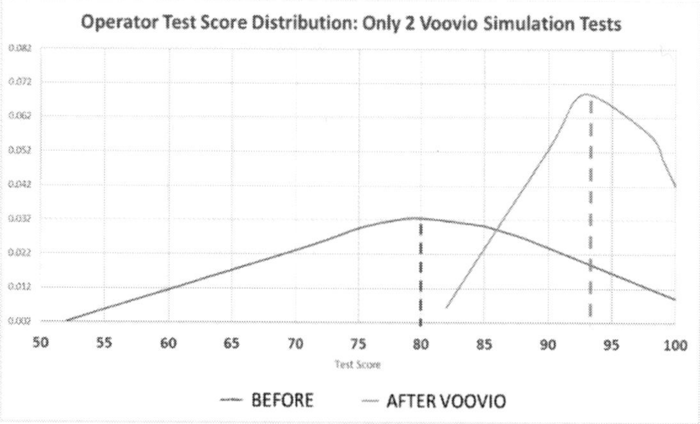

VOOVIO

2023 GLOBAL CONFERENCE
ON PROCESS SAFETY & BIG DATA
CCPS EPSC
OCTOBER 24 - 25, 2023
DECHEMA HOUSE | Frankfurt am Main, Germany
SUNCOR
CALUMET
Formosa Plastics
TESLA
Dow
PHILLIPS 66
VOOVIO
lyondellbasell
SHINTECH
BASF
ERGON
sabic
EASTMAN
FIBER INDUSTRIES
Chemours
Celanese
CRODA
plug
ALBEMARLE
TotalEnergies
BASF
HALLIBURTON
POLAND
GERMANY
FRANCE
ITALY
SPAIN
IRAN
JORDAN
ISRAEL
KUWAIT
BAHRAIN
QATAR
SAUDI ARABIA
UAE
OMAN
YEMEN

PETRONAS
DATA INNOVATION AT
PETRONAS
GAS PROCESSING KERTEH
VIA DATA CONTEXTUALIZATION
PRESENTERS
M Safwan M Diah ● Manager (Process & Operations Data), Gas Business, PETRONAS
Shahrulnizam Hanafi ● Manager (Data Management & Program), Gas Business, PETRONAS

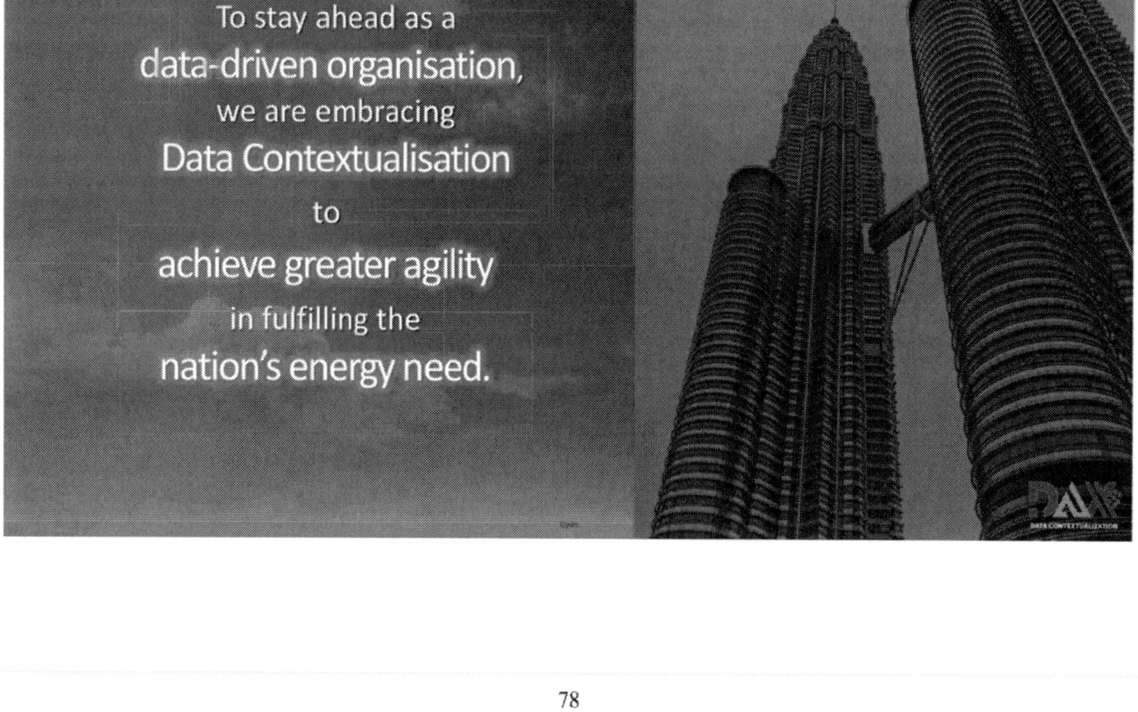
Reason
WHY
To stay ahead as a
data-driven organisation,
we are embracing
Data Contextualisation
to
achieve greater agility
in fulfilling the
nation's energy need.
DATA CONTEXTUALIZATION

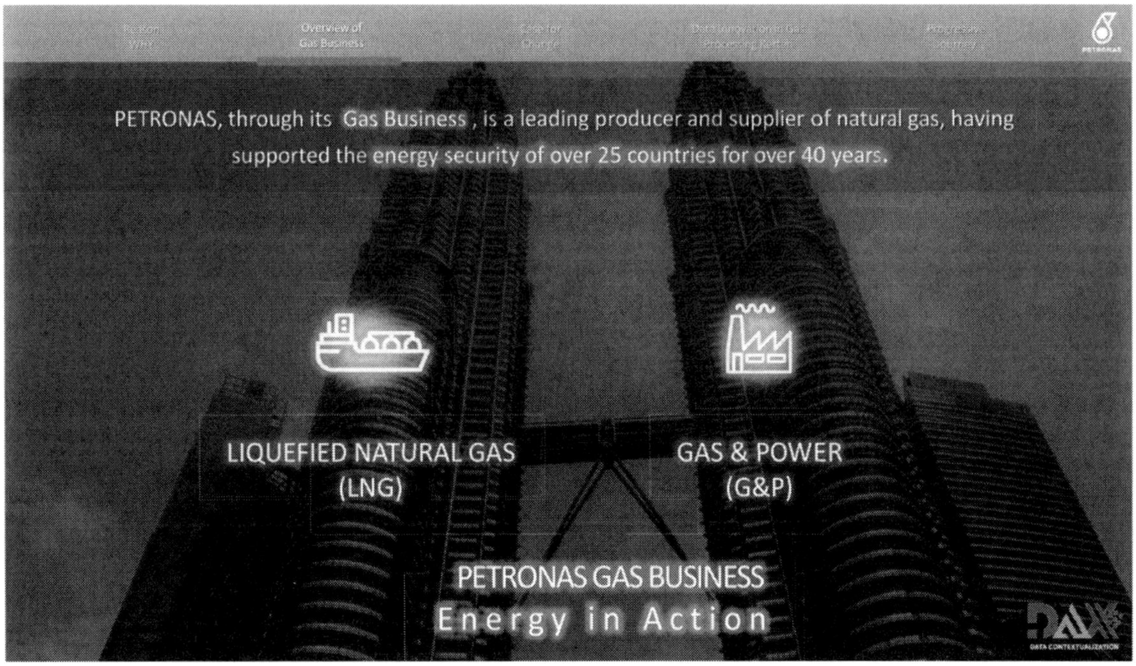
Overview of
Gas Business
PETRONAS, through its Gas Business , is a leading producer and supplier of natural gas, having supported the energy security of over 25 countries for over 40 years.
LIQUEFIED NATURAL GAS
(LNG)
GAS & POWER
(G&P)
PETRONAS GAS BUSINESS
Energy in Action

Case for
Change
OUR DATA INNOVATION STORY
GPK
GAS PROCESSING KERTEH

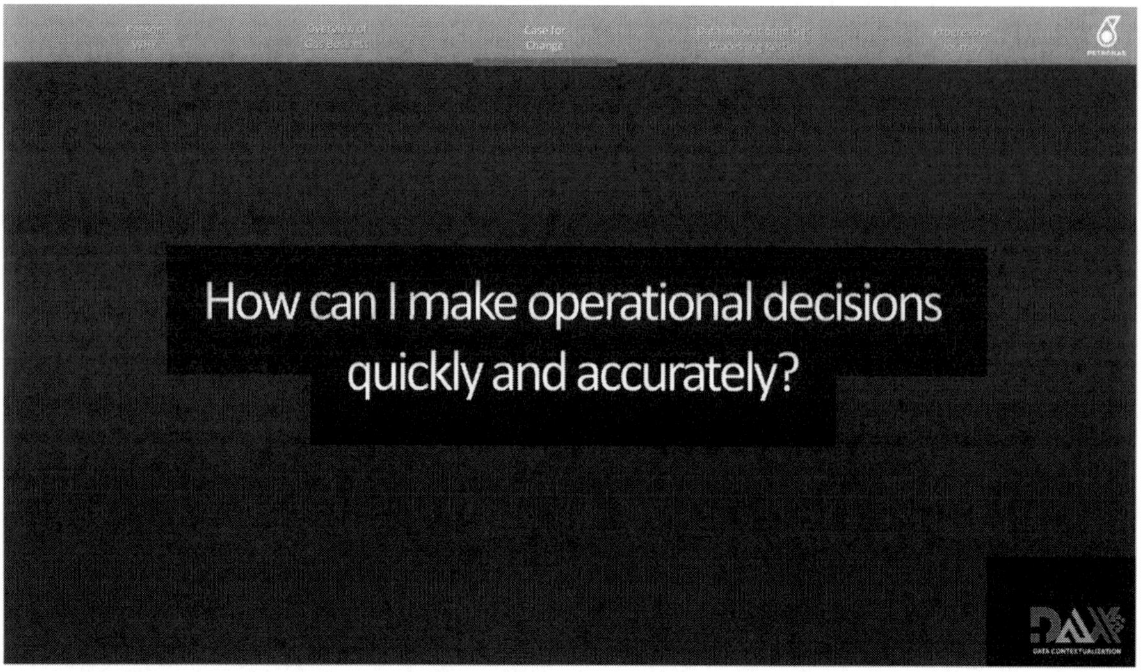
Case for
Change
How can I make operational decisions quickly and accurately?

Case for
Change
"As a process engineer, I am responsible in monitoring and optimizing our plant operations.
With that, I need real time plant data and related engineering drawings at the right time and manner."

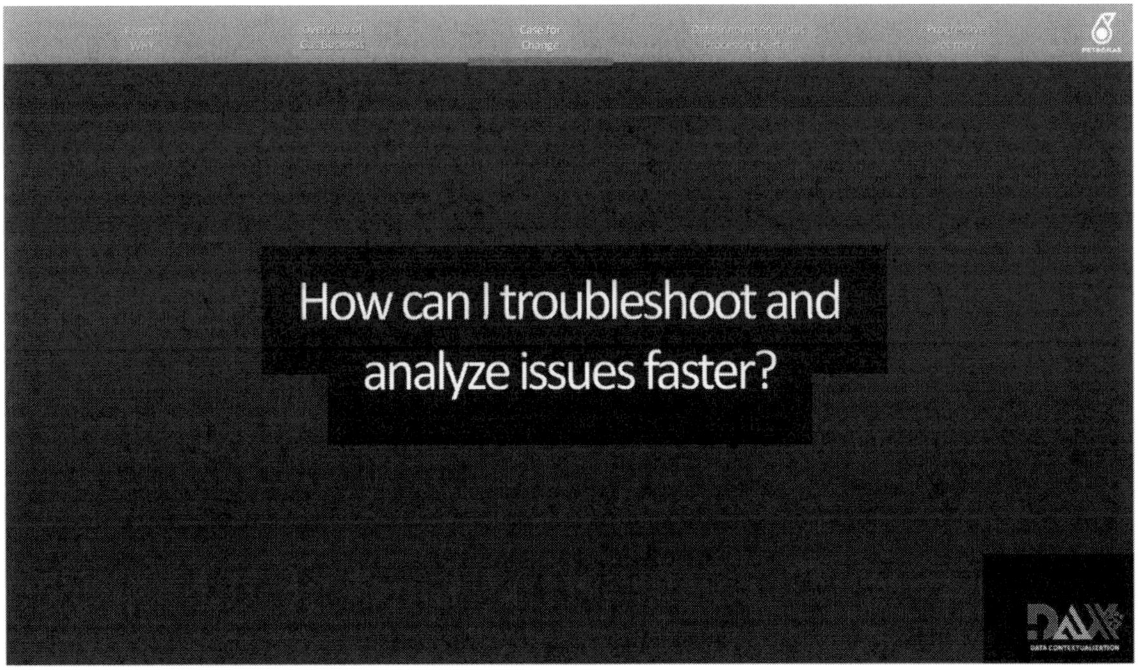
How can I troubleshoot and analyze issues faster?
DATA CONTEXTUALIZATION

"I'm a maintenance engineer and my job entails ensuring equipment and machinery are well maintained at all times throughout the plant.
To achieve this, I must perform regular inspections, troubleshooting, keeping records of all historical data and log reports to identify the gaps."
DATA CONTEXTUALIZATION

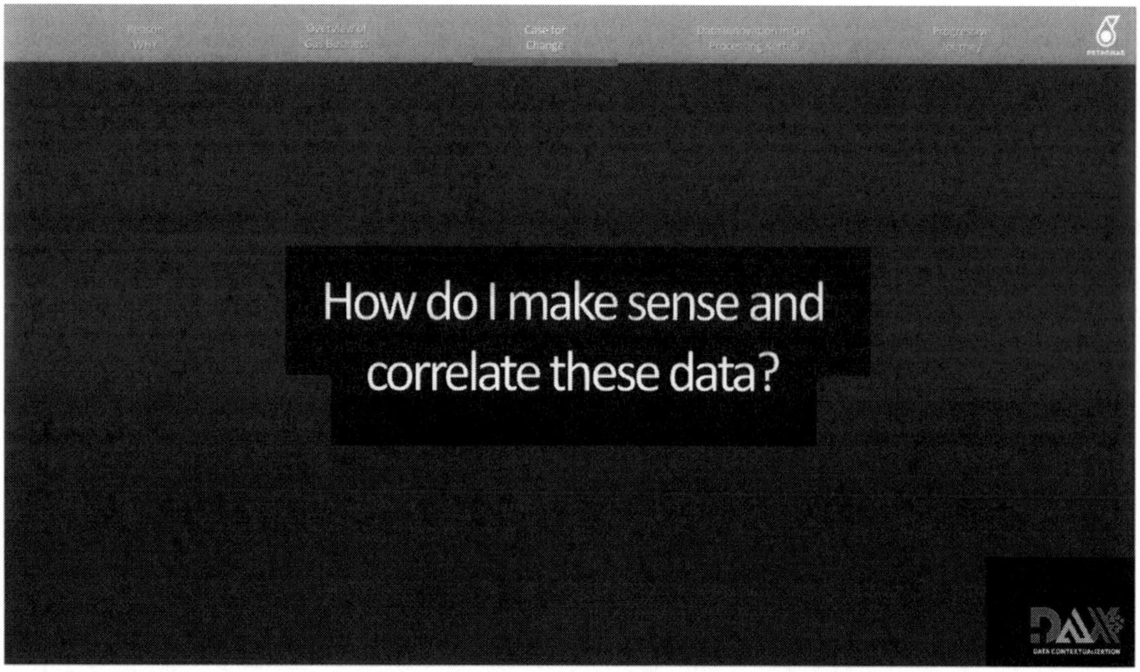
Case for
Change
How do I make sense and
correlate these data?

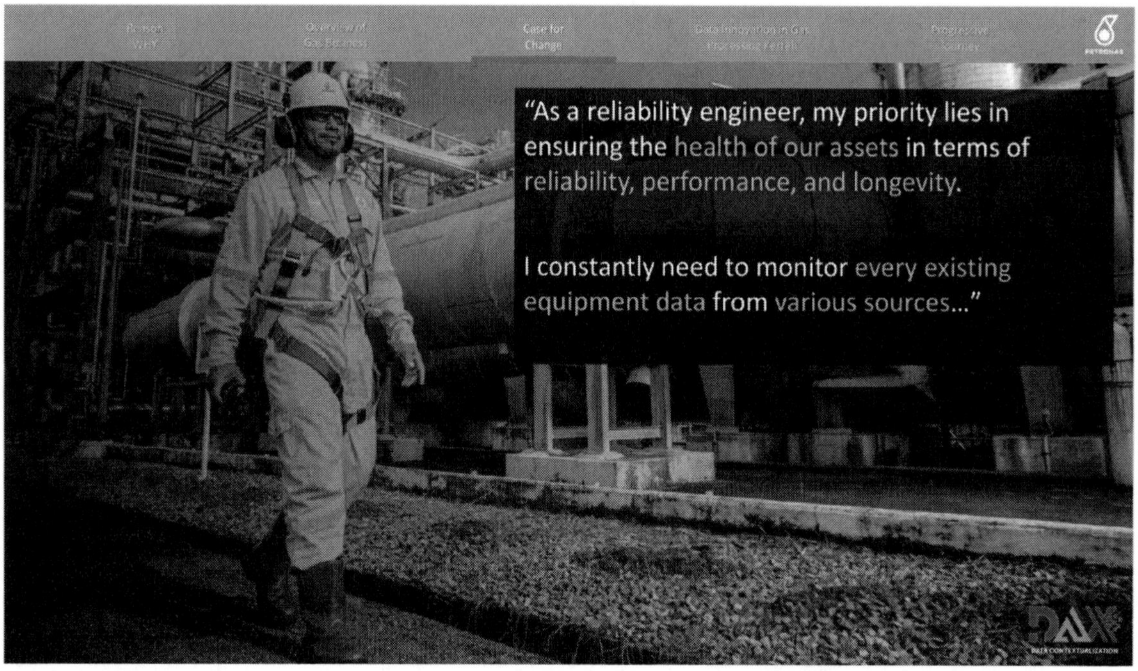
Case for
Change
"As a reliability engineer, my priority lies in ensuring the health of our assets in terms of reliability, performance, and longevity.
I constantly need to monitor every existing equipment data from various sources..."

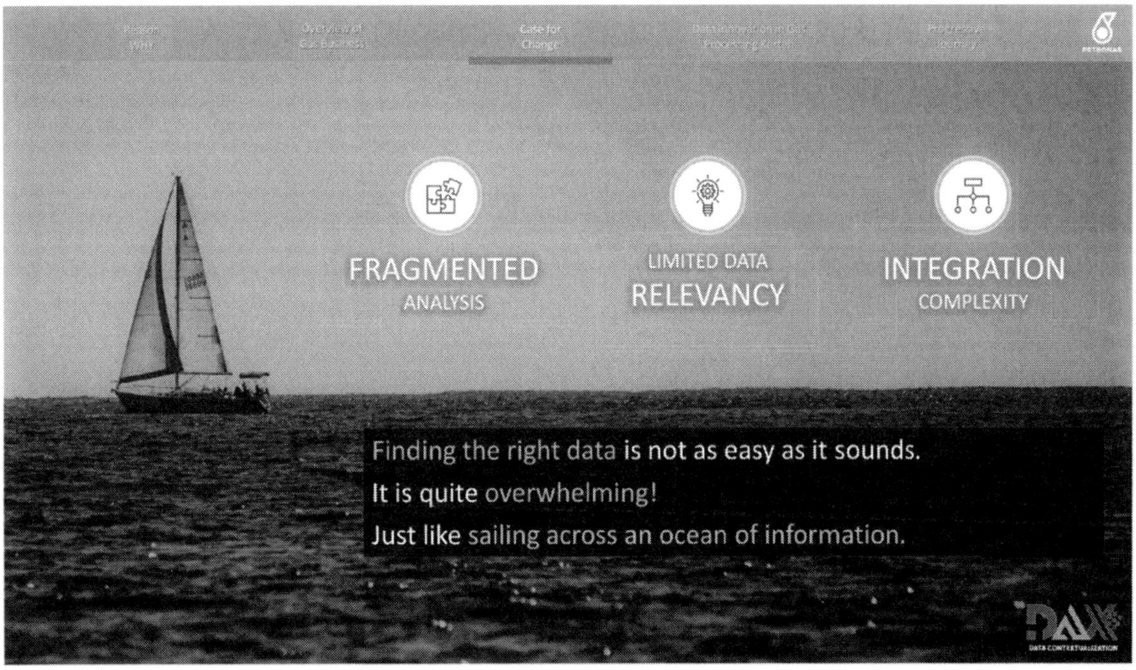
Case for Change
FRAGMENTED
ANALYSIS
LIMITED DATA
RELEVANCY
INTEGRATION
COMPLEXITY
Finding the right data is not as easy as it sounds.
It is quite overwhelming!
Just like sailing across an ocean of information.
DATA CONTEXTUALIZATION

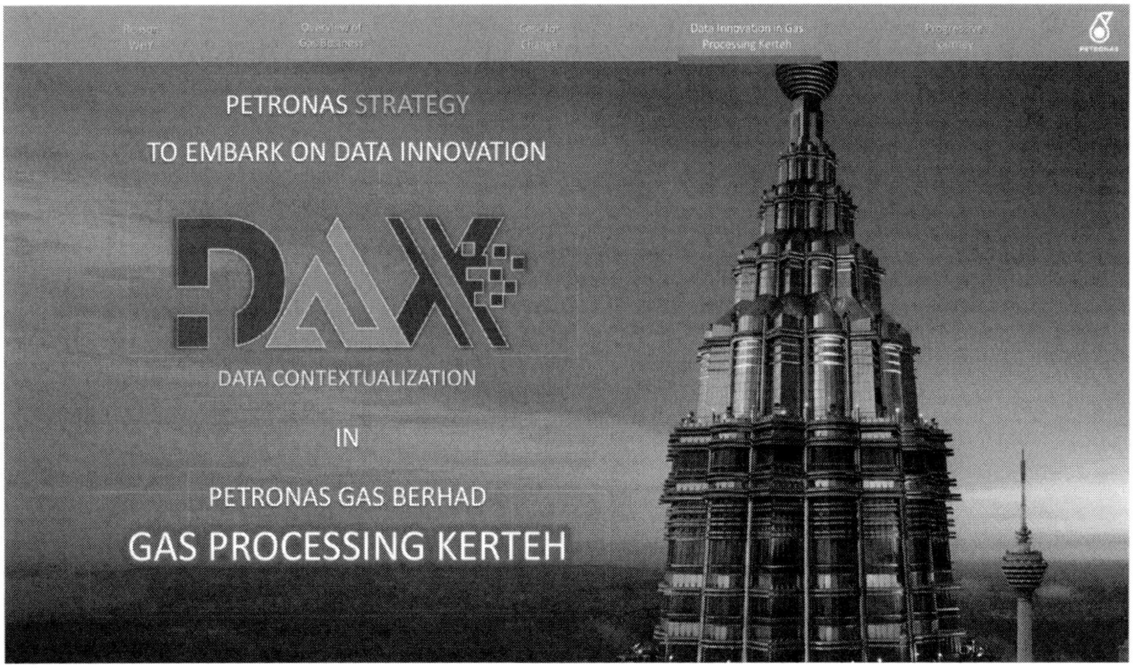
Data Innovation in Gas Processing Kerteh
PETRONAS STRATEGY
TO EMBARK ON DATA INNOVATION
DATA CONTEXTUALIZATION
IN
PETRONAS GAS BERHAD
GAS PROCESSING KERTEH

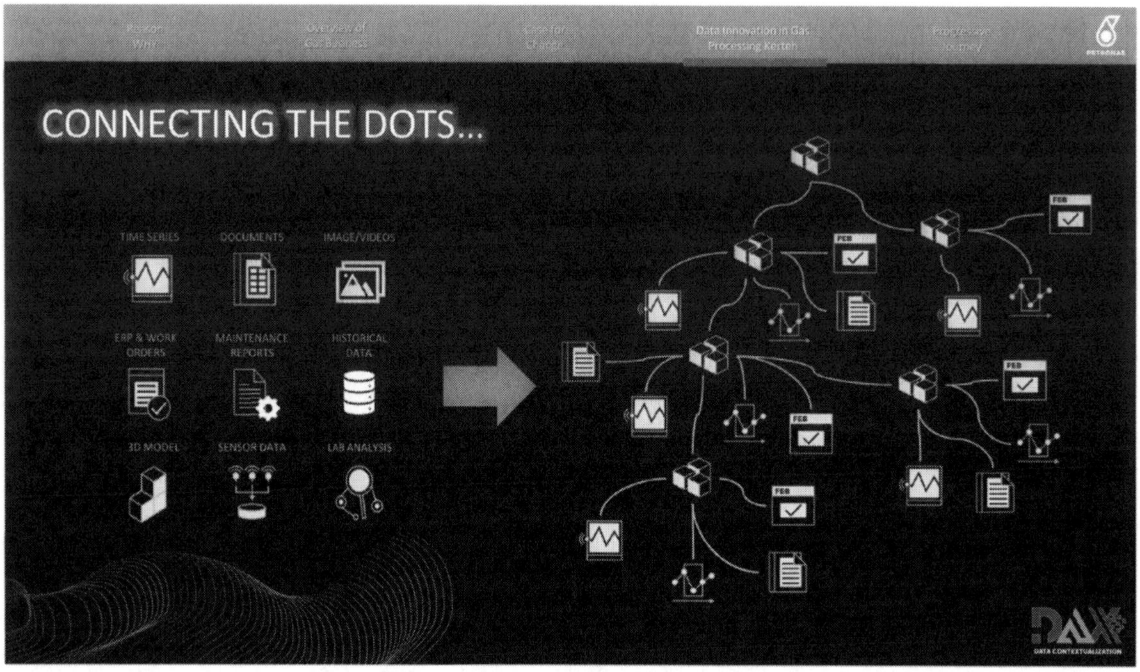
Data Innovation in Gas Processing Kerteh
CONNECTING THE DOTS...
TIME SERIES
DOCUMENTS
IMAGE/VIDEOS
ERP & WORK ORDERS
MAINTENANCE REPORTS
HISTORICAL DATA
3D MODEL
SENSOR DATA
LAB ANALYSIS
FEB
DATA CONTEXTUALIZATION

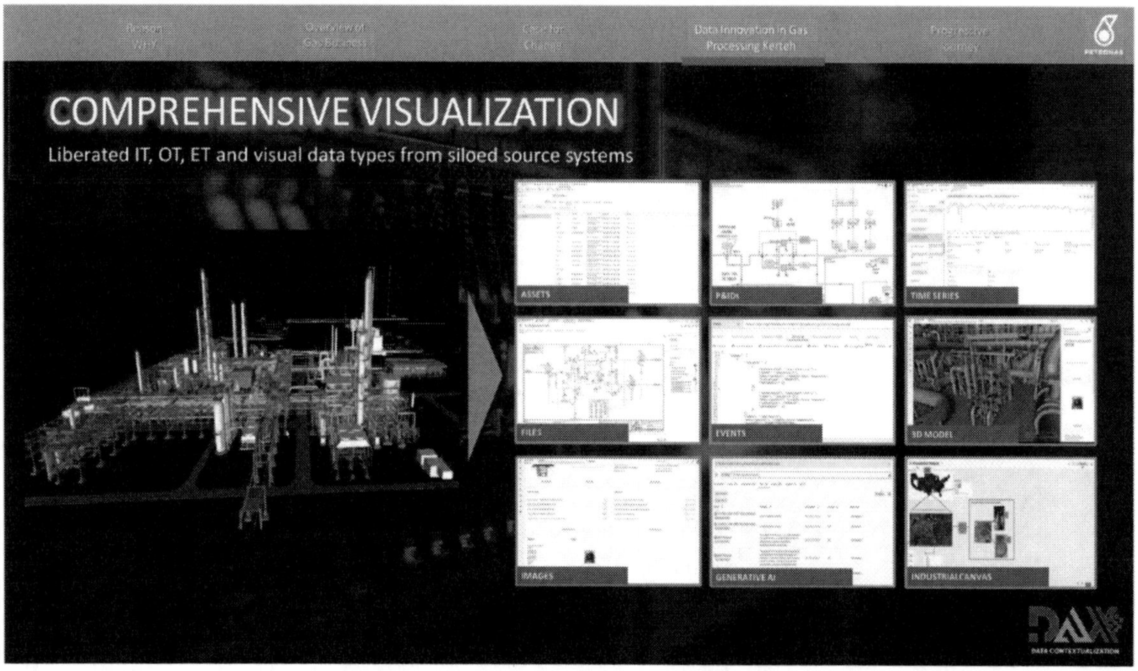
Data Innovation in Gas Processing Kerteh
COMPREHENSIVE VISUALIZATION
Liberated IT, OT, ET and visual data types from siloed source systems
ASSETS
P&IDs
TIME SERIES
FILES
EVENTS
3D MODEL
IMAGES
GENERATIVE AI
INDUSTRIALCANVAS
DATA CONTEXTUALIZATION

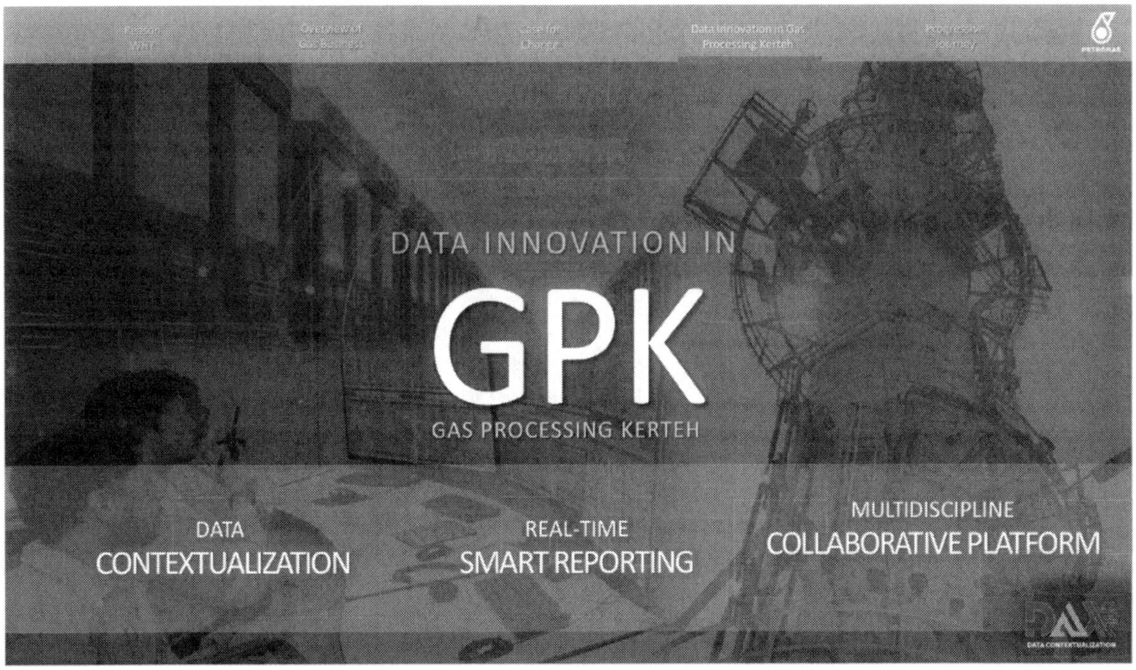
Data Innovation in Gas Processing Kerteh
DATA INNOVATION IN
GPK
GAS PROCESSING KERTEH
DATA
CONTEXTUALIZATION
REAL-TIME
SMART REPORTING
MULTIDISCIPLINE
COLLABORATIVE PLATFORM

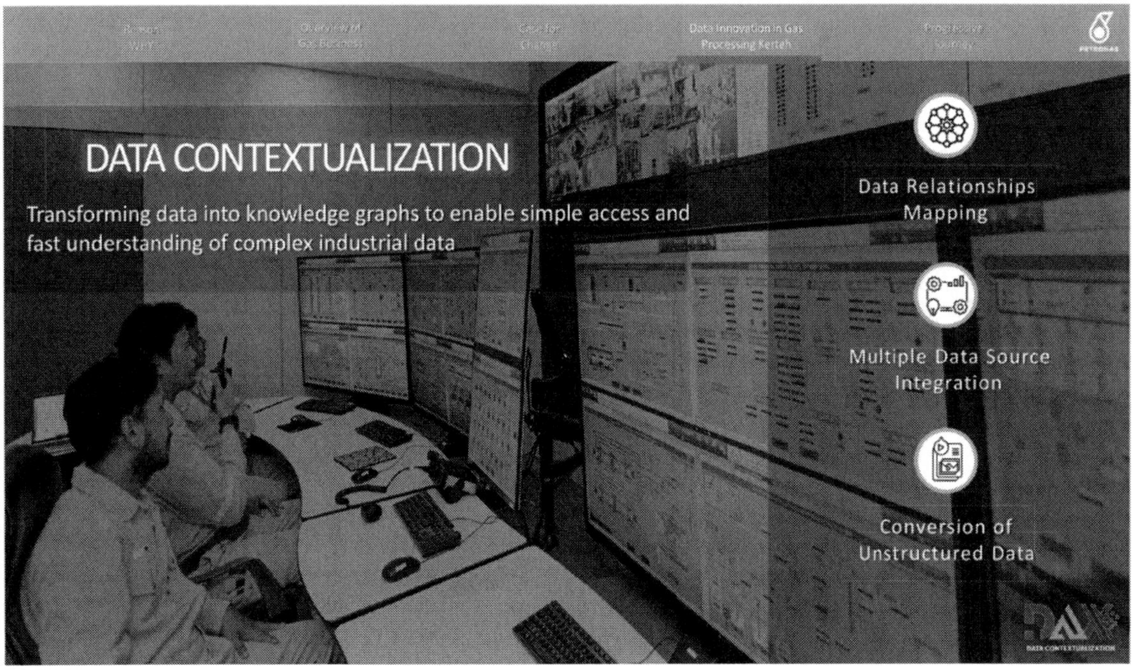
Data Innovation in Gas Processing Kerteh
DATA CONTEXTUALIZATION
Transforming data into knowledge graphs to enable simple access and fast understanding of complex industrial data
Data Relationships Mapping
Multiple Data Source Integration
Conversion of Unstructured Data

Data Innovation in Gas
Processing Kerteh
REAL-TIME SMART REPORTING
Empowering prompt and precise troubleshooting for operational efficiency
Anomalies Detection
Alerts & Notifications
Personalized Reporting
Efficient Data Retrieval
Proactive Response & Decision-Making

Data Innovation in Gas
Processing Kerteh
MULTIDISCIPLINE
COLLABORATIVE PLATFORM
Reinforcing simultaneous collaboration across disciplines
Centralized Digital Workspace
Cross-Functional Dashboards
Industrial Canvas

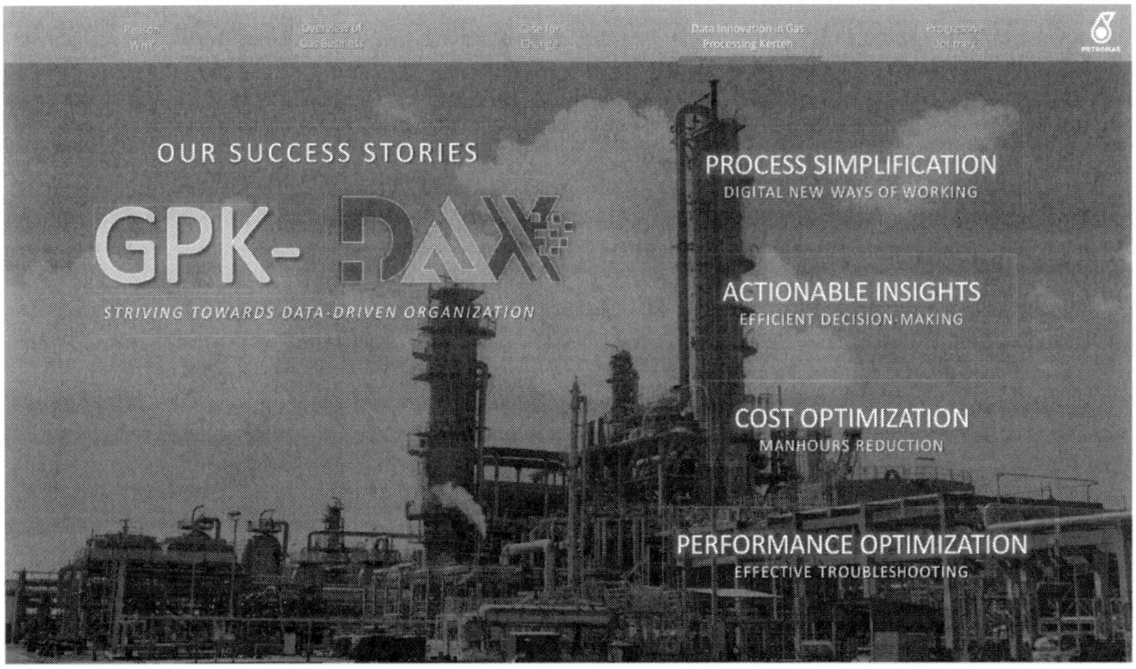
Data Innovation in Gas Processing Kerteh
OUR SUCCESS STORIES
GPK- DAX
STRIVING TOWARDS DATA-DRIVEN ORGANIZATION
PROCESS SIMPLIFICATION
DIGITAL NEW WAYS OF WORKING
ACTIONABLE INSIGHTS
EFFICIENT DECISION-MAKING
COST OPTIMIZATION
MANHOURS REDUCTION
PERFORMANCE OPTIMIZATION
EFFECTIVE TROUBLESHOOTING

Progressive Journey
GPK- DAX
This strategic initiative will pave the way for digital transformations to take place at other assets across Gas Business and PETRONAS

THANK YOU
CONTACT DETAILS
M Safwan M Diah
Manager (Process & Operations Data)
Gas Business, PETRONAS
msafwan.mdiah@petronas.com.my
Shahrulnizam Hanafi
Manager (Data Management & Program)
Gas Business, PETRONAS
shahrulnizam.hanaf@petronas.com.my

leading to win

October 24, 2023

Enhancing Reliability with Data Science

Energy lives here

Paul Kenny, Reliability Analytics Advisor, Global Operations & Sustainability, ExxonMobil
Nyala Noë, Lead Data Scientist, Empirisys

we are ExxonMobil

Who we are

ExxonMobil (EM), one of the largest publicly traded international energy and petrochemical companies, creates solutions that improve quality of life and meet society's evolving needs.

The corporation's **three primary businesses** - Upstream, Product Solutions and Low Carbon Solutions - provide products that enable modern life, including energy, chemicals, lubricants, and lower-emissions technologies.

leading to win

2

Challenge to address

Enhancing Reliability of an Existing Key Asset with Data Science.

Challenge: the management team overseeing the asset performs continuous improvement to:

- Safety
- Health and Environment
- Reliability
- Production

In addition to the existing and well-established approaches to successfully manage these areas, can the latest innovations in data science also be applied to yield even greater outcomes?

leading to win

3

Outcomes

Asset reliability improvements evident post-intervention

Utilising data science, we identified supporting data evidence to pinpoint areas requiring improvement. Subsequently, a targeted plan was developed and executed to address these areas. The high-value items were quickly implemented, and reliability has improved significantly since last year.

1. Number of unplanned outages reduced by over half.
2. Improvements in several key systems
3. Engagement of the whole operating team, setting them up for success

leading to win

4

A partnership for success

We are focused on establishing ourselves as industry leaders with our operating performance. The application of **Data Science** methods to **improve operating performance** is a relatively new and rapidly evolving area. We therefore wished to explore if we could continue to evolve our approach through the application of cutting-edge industry technology.

The local asset management has extensive experience in operating it successfully. It has a strong and diverse workforce as well as a wide range of accumulated operating data. Could Data Science enhance our operating decisions?

Empirisys' belief that changing culture is fundamental to creating the conditions that enable safer, better performing, more environmentally responsible and, ultimately, more sustainable businesses to evolve. These goals can be achieved with a more data driven approach in which Empirisys excel.

leading to win

5

Approach

Working as one team to respond to the challenge.

The combined team of Empirisys and EM used an Agile approach to the challenge posed to be able to pivot quickly depending on the results of the data analysis. The approach proved to be a success, with the team being able to quickly respond to new hypotheses and questions being generated as insight was getting uncovered.

1. Weekly stand-ups to share progress and change direction if needed
2. Iterative analysis that could react quickly to new questions as they emerged
3. Regular reviews with Subject-Matter Experts to make sure data insights make sense

leading to win

6

Approach

Working as one team to respond to the challenge.

The roles in the team, Full Time (FT) or Part Time (PT):

- Project Leader (FT): Ensure the smooth running of the project and connecting the right people
- Senior Managers (PT): Validate results and action them
- Subject-Matter Experts (PT): Interpret, validate, and review results
- Data Scientists (FT): gather, prepare, and analyse data
- SCRUM Master (PT): facilitate the team working together

leading to win

7

Approach

1. Set objective: what problem needs to be addressed?
Example: Understand the effect of operator experience on asset reliability

2. Gather data: gather any data relevant to the objective
Example: reliability and training/competence data of asset personnel

3. Analyse evidence: analyse the gathered data for patterns and anomalies
Example: trends in experience levels over time

4. Gain insight: root cause analysis of trend
Example: small but consistent experience drift over time

5. Derive action: use data to derive a meaningful solution
Example: focused effort on coaching

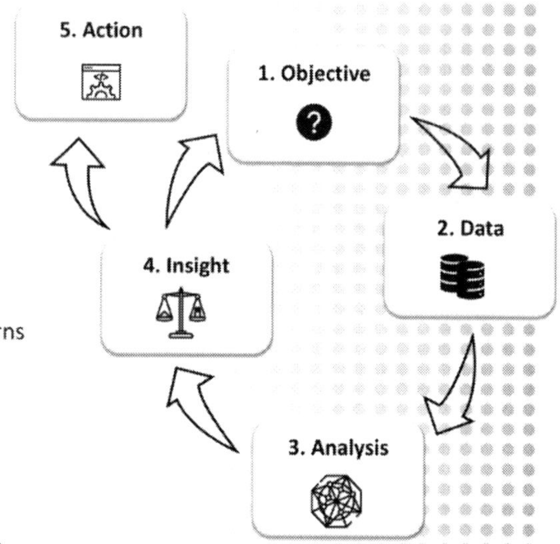

The loop can be repeated as often as needed to get to an actionable insight

leading to win

8

Combining different datasets allows us to build up a complete picture

What is the current state of the asset and its people?	What is the history of the asset & how did that influence its current state?	How do the people feel about their asset and its current state?
Operational data	Historic data	
▪ Plant	▪ Trends	▪ Areas of strength
▪ Process	▪ Anomalies	▪ Areas of opportunity
▪ People	▪ Forecasting	▪ Emotions

It is the combination of these data sources and their correlations that allowed the team to pull out meaningful insight. In isolation, each dataset only offered a part of the puzzle.

Actioning the findings - example

Details on Intervention: rationalize operator surveillance processes.

Approach: SME team reviewed the value of the current list of procedures.

Result: one third of procedures removed, allowing more time for other activities.

Reduction in the Number of Procedures Completed

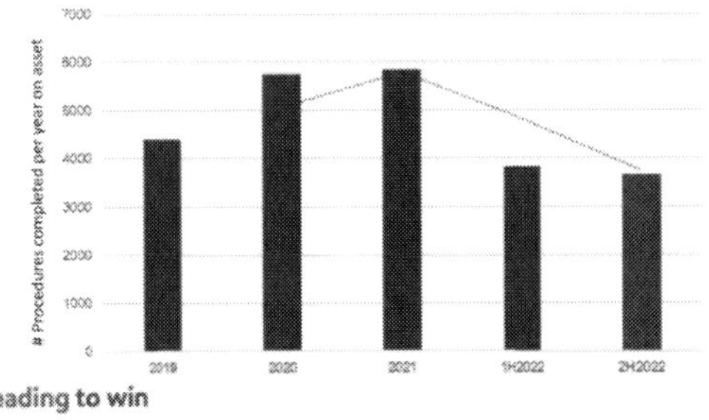

Increased Procedure Duration Maintained

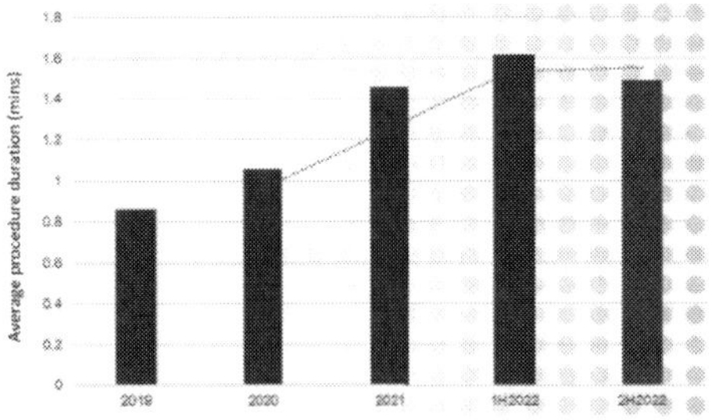

Asset Outcome

Reliability outcomes for the asset improved as the findings were completed in 2Q2022 onwards

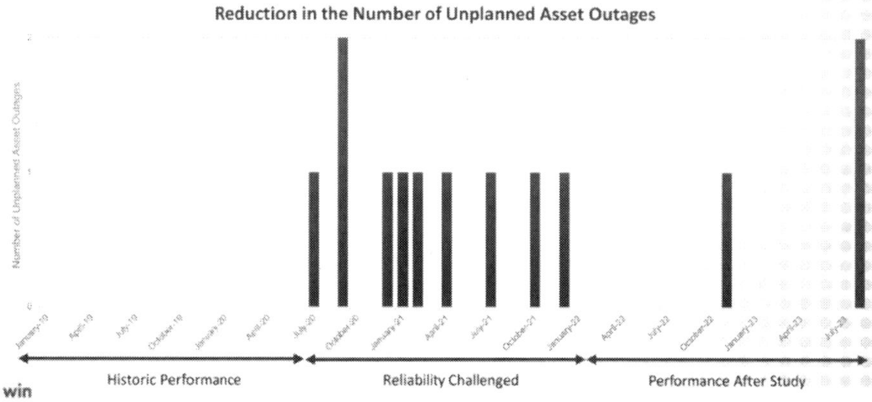

leading to win

11

Winning Factors

Seamless collaboration and implementing actions informed by data evidence ensured success.

- Strong support from asset manager
- Team collaboration and transparency
- Efficient and wide-ranging data access
- Data science platform that enabled quick analysis
- Frequent touchpoints with relevant subject-matter experts
- Intervention to directly address the outcomes of the analysis

leading to win

12

Looking to the Future

- Data science is moving quickly, need to **constantly** stay in touch to be at the **cutting edge**, use experts collaboratively to achieve this
- Continue to seek **strong data evidence** before making decisions
- Data science has the potential for more **proactive approaches**, not just reactive

1	2	3
Be thought leaders in using data science in industry	Data-driven decision making at the heart of everything we do	Proactively identify and address issues as they arise

leading to win

13

leading to win

October 24, 2023

Thank you

Energy lives here

we are ExxonMobil

Operationalising Big Data & Digitalisation to Improve Process Safety Performance

INDUSTRY PANEL DISCUSSION

24TH OCTOBER 2023

2023 GLOBAL CONFERENCE ON PROCESS SAFETY & BIG DATA

CCPS EPSC

Previous Control Room

Control Room Today

Data - How to take profit of Data for the Safety ?

- More data is being created than ever before

~50%* of existing industrial data as of 2021 was created in the two preceding years

- A big part of all digital data is unstructured
- Yet not knowing the best way to read, understand, and apply data

For example, Approximately 68%** of currently available enterprise data goes unused because it sits in organizational silos

- Digitalisation : A long journey

Data collection→ Data structuration → Data Contextualization → Data Analytics → Data Enhancement → DIGITALISATION

*Statista, ** Seagate's "Rethink Data" report

How might we make the best of our existing Data

To detect at an early stage both occupational and major risks and prevent accidents?

To increase the skills of employees/operators on safety issues ?

To ease access to relevant knowledge & experience for Safety?

To make sure that all entities and sites have a consistent risk assessment ?

To visualize and understand all the risks and corrective action plans to manage the risks?

To better monitor the integrity of our installation; maintenance, inspection, tests, ...?

To ensure the status of the barriers (availability, efficiency,...)?

To . . .

Discussion

Thank You

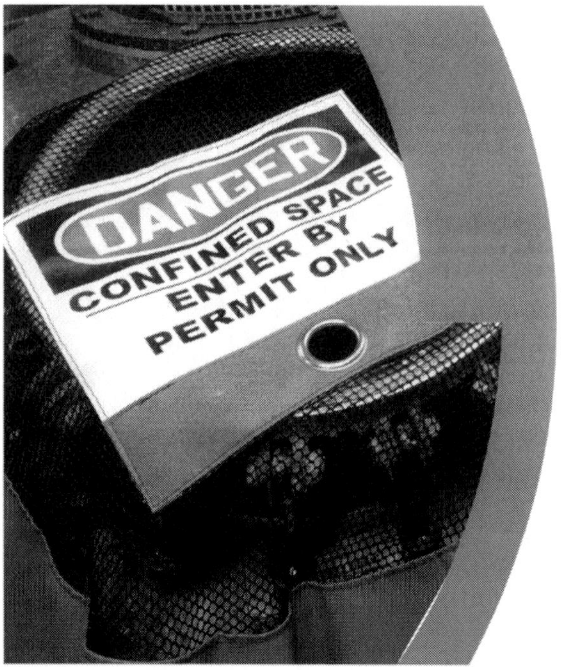

covestro

Robots Can Minimize Confined Space Entries of Pressurized Vessels / Reactors

by Marcel Swalen

Short introduction

covestro

Marcel Swalen
Reliability Engineer

Covestro WAALWIJK The Netherlands

- Site Waalwijk: 60 years anniversary in 2022
- Our products
  Specialty coatings resins:
  - poly-acrylates as powder (beads) and water-based (emulsions)
  - poly-urethanes (water-based)
- End markets
  Printing, packaging, industrial coatings, decorative paint and coatings, flooring
- ~500 employees

Agenda

1. Legislation
2. Typical inspection
3. Camera inspection

Legislation of Pressure equipment

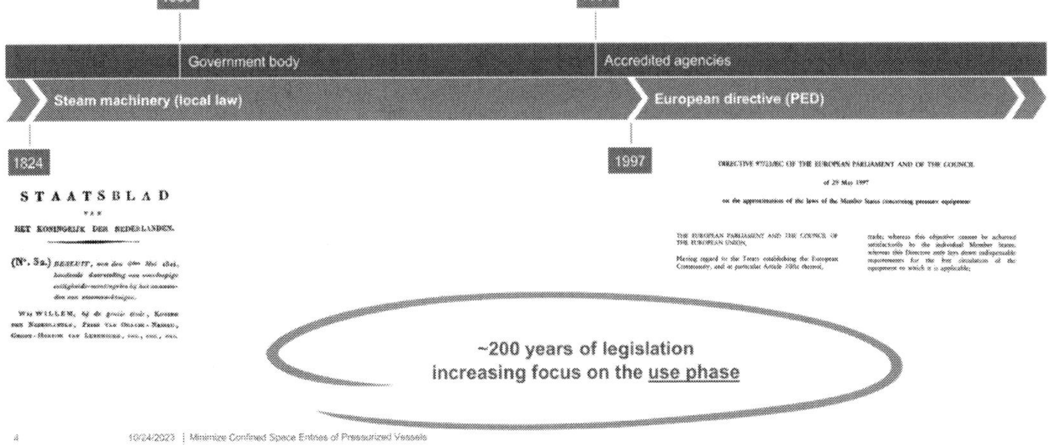

Typical inspection

covestro

Preparation	Execution	Evaluation
Inspection plan agreed upfront with notified body: • Visual inspection by entering the pressure vessel	Inspection on-site: • Emptying and cleaning vessel • Disconnect all piping • Building scaffolding • Work permit for confined space entry • Various safety measures: – Gas monitor – Rescue personnel standby – Rope access – Halting nearby equipment to clear ATEX zone	Report by notified body: • Approval for another 4 or 6 years

Camera inspection

covestro

Preparation	Execution	Evaluation
Inspection plan agreed upfront with notified body: • Visual inspection by <ins>**camera**</ins> • **Review of possible degradation scenarios** • **Procedure on camera equipment** • **Inspector reserves right to enter vessel**	Inspection on-site: • Emptying and cleaning vessel ~~• Disconnect all piping~~ ~~• Building scaffolding~~ ~~• Work permit for confined space entry~~ ~~• Various safety measures:~~ ~~– Gas monitor~~ ~~– Rescue personnel standby~~ ~~– Rope access~~ ~~– Halting nearby equipment to clear ATEX zone~~	Report by notified body: • Approval for another 4 or 6 years

Steam vessel (pilot 2021)

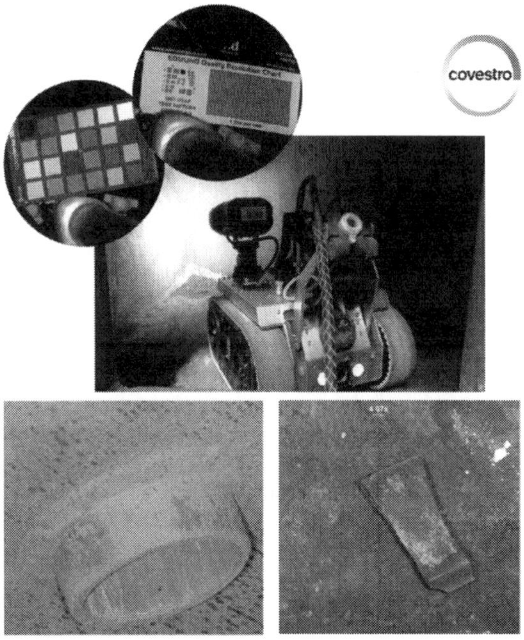

Reactor vessel (inspection 2022)

EX-camera

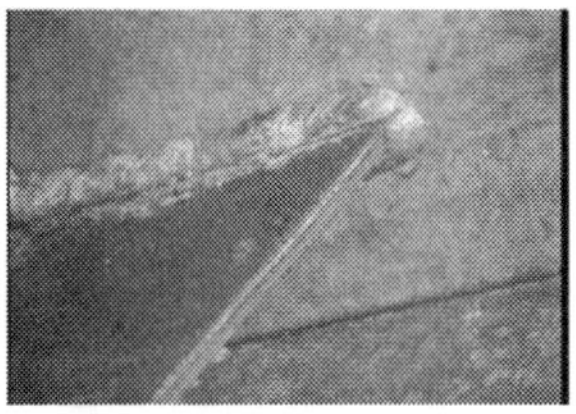

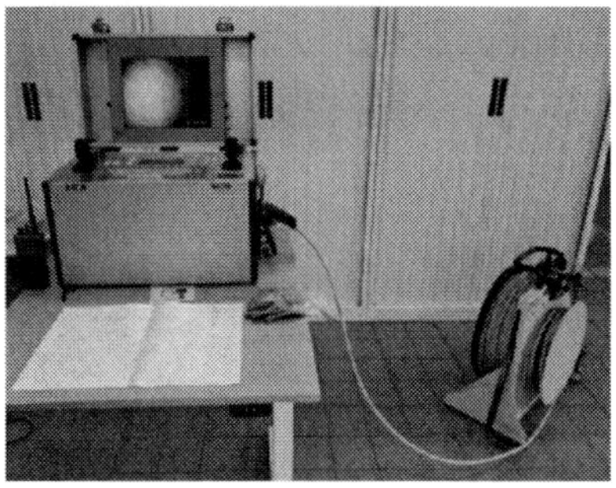

Reactor vessel (inspection 2022)

non-Ex camera

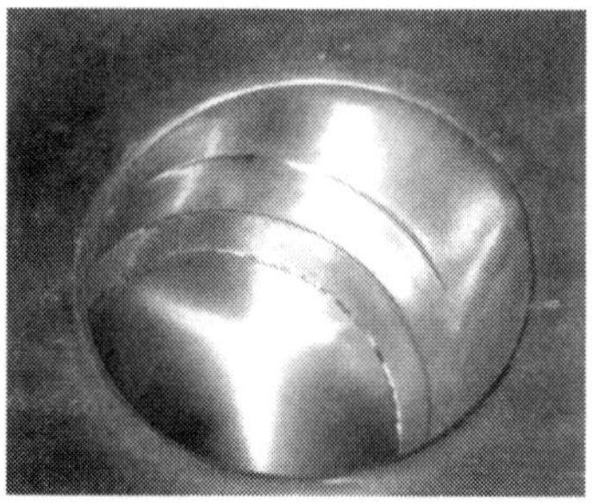

9 10/24/2023 | Minimize Confined Space Entries of Pressurized Vessels

Takeaways

Camera

Ex and non-Ex versions

Good image quality

Many configurations (drone, crawler)

Compliance

Camera is no guarantee! (just like any other tool)

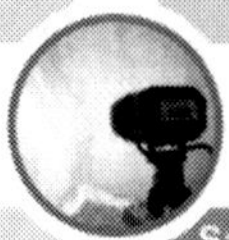

Methodology

More work upfront

Less impact on production

Safety

Reducing confined space entries!

Only enter vessel when necessary

10 10/24/2023 | Minimize Confined Space Entries of Pressurized Vessels

2023 GLOBAL CONFERENCE ON PROCESS SAFETY & BIG DATA

CCPS EPSC

2023 GLOBAL CONFERENCE ON PROCESS SAFETY & BIG DATA

CCPS EPSC

Data Migration to Cloud for Consistent/Quality PHA

Speaker Profile

- Edward M. Marszal, President and CEO of Kenexis Since 2004
  - Technical safety consultancy and software developer
  - BSChE, Ohio State University, 1992
- Author
  - SIL Selection with Layer of Protection Analysis, Security PHA Review
- Presentation Team
  - Matthieu Tagliaferri and Romain Raballand – Yara
  - Sean Cunningham – Kenexis

2023 GLOBAL CONFERENCE ON PROCESS SAFETY & BIG DATA
CCPS EPSC
OCTOBER 24 - 25, 2023
DECHEMA HOUSE | Frankfurt am Main, Germany

Introduction

- Large investments are made in performing PHA
- Data generated has value beyond initial task
  - Risk registers, safeguard performance targets, stakeholder communication
- Enterprise functionality desired
  - Access to global data, summary data/dashboarding, design automation
- Case study of migration to cloud-based enterprise system

Problem

- Large number of PHA studies scattered around the world
  - Excel Files - Templates
  - Inconsistent, No visibility, Insecure
- Inconsistent, manual handling of SIS design basis
- No tools to carry data to subsequent tasks
- Corporate knowledge lost, not carried to other organizational areas and inconsistent results

2023 GLOBAL CONFERENCE ON PROCESS SAFETY & BIG DATA
CCPS EPSC
OCTOBER 24 - 25, 2023
DECHEMA HOUSE | Frankfurt am Main, Germany

Extension of Partial Solution

- Extensive templating deployed to address PHA inefficiency
  - Lessons learned distributed through organization
  - Inconsistent valuation of causes and safeguards in LOPA
  - Unit-to-unit and site-to-site inconsistency in analysis for identical design
  - Inefficiency of starting from "blank sheet of paper" for proven designs
- Deployment of templating through spreadsheet templates sub-optimal

Problem Analysis for Solution

- Cloud-based system will allow global access to all data
- Unified data structure for consistency - dashboarding and data export
- Advanced features for templating and libraries
- Suite of integrated components to allow for data handoff to other tasks
- Consistent implementation of SIS design – SIL calculation and SRS
- Use of software both online in the cloud and offline when remote

2023 GLOBAL CONFERENCE ON PROCESS SAFETY & BIG DATA
CCPS EPSC
OCTOBER 24 - 25, 2023
DECHEMA HOUSE | Frankfurt am Main, Germany

Requirements for Solution

- Data security and privacy, single sign-on
- EU data privacy laws
- Migration of existing data to cloud-based solution

2023 GLOBAL CONFERENCE ON PROCESS SAFETY & BIG DATA

CCPS EPSC

OCTOBER 24 - 25, 2023
DECHEMA HOUSE | Frankfurt am Main, Germany

Solution

- Kenexis Integrated Safety Suite
  - HAZOP/LOPA in cloud and desktop
  - Unified data structure using open formats (JSON)
  - Integrated SIS Safety Lifecycle Management (including FTA tool)
- Not all desired functionality available
  - List of required functionality
  - Commercial requirement to implement
- Pilot process to test efficacy

2023 GLOBAL CONFERENCE ON PROCESS SAFETY & BIG DATA

CCPS EPSC

OCTOBER 24 - 25, 2023
DECHEMA HOUSE | Frankfurt am Main, Germany

Migration Process

- Primarily Excel spreadsheets
- Tool developed to allow users to convert Excel files to JSON object which is opened on desktop or uploaded to cloud
- Multiple versions of Excel files required to be handled as multiple templates in use

Commercial Results

- Pilot study successful
- Software deployed to pilot sites around the world
- All promised software features delivered
- Migration of data by end user successful
- Full commercial deployment

2023 GLOBAL CONFERENCE
ON PROCESS SAFETY & BIG DATA
CCPS EPSC
OCTOBER 24 - 25, 2023
DECHEMA HOUSE | Frankfurt am Main, Germany

Future Work

- Enhanced and customizable dash-boarding capabilities
- Export/import to “in suite” bow-tie diagram application
- API for integration of data with ERP and maintenance applications

Summary

- International chemical company desired cloud-based PHA software for global consistency and quality
- Cloud-based solution implemented and customized for purpose
- Total data migration implemented, primarily by end user with tools
- Successful pilot leads to full commercial deployment

2023 GLOBAL CONFERENCE
ON PROCESS SAFETY & BIG DATA

CCPS EPSC

2023 GLOBAL CONFERENCE
ON PROCESS SAFETY & BIG DATA

CCPS EPSC

Technology Panel

on

Current Technology Developments for Digital Solutions in Process Safety Applications

Co-organized by CCPS & EPSC

2023 GLOBAL CONFERENCE ON PROCESS SAFETY & BIG DATA

CCPS EPSC

OCTOBER 24 - 25, 2023
DECHEMA HOUSE | Frankfurt am Main, Germany

Arturo Trujillo, Global Service Group Manager-Process Safety, Dekra

- 35+ years of experience as a process safety practitioner.
- Professional career developed in consulting and EPC companies.
- Currently, leading the process safety digitalization program in DEKRA.

Co-organized by CCPS & EPSC

2023 GLOBAL CONFERENCE ON PROCESS SAFETY & BIG DATA

CCPS EPSC

OCTOBER 24 - 25, 2023
DECHEMA HOUSE | Frankfurt am Main, Germany

Gus Carroll, CEO and co-founder of Empirisys

- 37+ years of experience as a process engineering, process safety and risk management.
- Current focus is developing data science applications to address safety, reliability, and performance problems for industrial sector.
- Holds BEng in Chemical Engineering and MSc Process Engineering.

Co-organized by CCPS & EPSC

2023 GLOBAL CONFERENCE ON PROCESS SAFETY & BIG DATA

CCPS EPSC

OCTOBER 24 - 25, 2023
DECHEMA HOUSE | Frankfurt am Main, Germany

Benjamin Lee, Product Owner – Digital Programs, Siemens Energy

- Currently - Product Owner, Digital Programs, Siemens Energy, *Erlangen Germany.*
- Consultant, Siemens Corporate Technology, *Munich Germany.*
- Focus on Decision Support Systems for Uncertainty Management in Offshore, Pulp & Paper, & Energy Systems.
- Holds PhD Mechanical Engineering, Georgia Institute of Technology, *Atlanta USA.*

Co-organized by CCPS & EPSC

Are we going too fast?

What's in it for me?

Key Issues

- Garbage in, garbage out
- Use it all, not just the easy stuff
- Who is it for?....and why?
- When reality bites!

empirisys

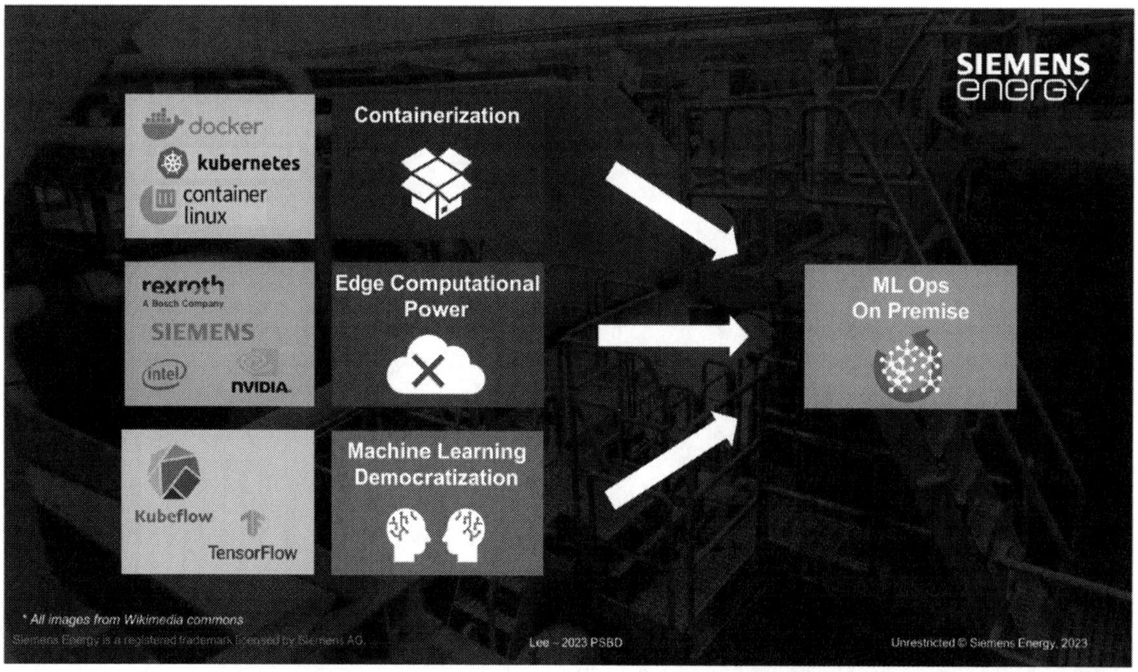

2023 GLOBAL CONFERENCE ON PROCESS SAFETY & BIG DATA
CCPS EPSC
OCTOBER 24 - 25, 2023
DECHEMA HOUSE | Frankfurt am Main, Germany
Current Technology Developments for Digital Solutions in Process Safety Applications
Panelist
Arturo Trujillo
Global Service Group Manager-Process Safety
DEKRA
Panelist
Gus Carroll
CEO and co-founder
empirisys
Panelist
Dr. Benjamin Lee
Product Owner - Digital Programs
SIEMENS energy
Moderator
Dr. Abdul Aldeeb
Process Safety Portfolio Manager
SIEMENS energy
Co-organized by CCPS & EPSC

2023 GLOBAL CONFERENCE
ON PROCESS SAFETY & BIG DATA

CCPS EPSC

2023 GLOBAL CONFERENCE
ON PROCESS SAFETY & BIG DATA

CCPS EPSC

Exploratory approach to develop a mathematical model using neural networks to estimate potential damages after process safety events in the oil and gas facilities.

Oscar Mauricio Barajas Pinzon
Loss Prevention Engineer
CSP, ASP, CFPS, ISO 45001, NEBOSH IGC.

2023 GLOBAL CONFERENCE ON PROCESS SAFETY & BIG DATA

CCPS EPSC

OCTOBER 24 - 25, 2023
DECHEMA HOUSE | Frankfurt am Main, Germany

Oscar M. Barajas Pinzon

- Loss Prevention Engineer
- (7 Years in Saudi Aramco)
- Abqaiq LPD Division.
- 20 years of experience in Oil & Gas Industry in Colombia, Peru, Venezuela, USA and Saudi Arabia.
- BSc in Mechanical Engineering.
- MSc in Industrial automation and Fire Protection Engineering.
- Certifications: CSP, CFPS, NEBOSH IGC,IFC,IOG,PSM. ISO 45001.

Use of Neural Nets for the calculation of potential damages in oil and gas facilities affected by Process Safety Event:

Type of process safety event.

Type of process.

Number of workers.

Number of contractors.

Presence of nearby communities.

Safety Instrumented Systems.

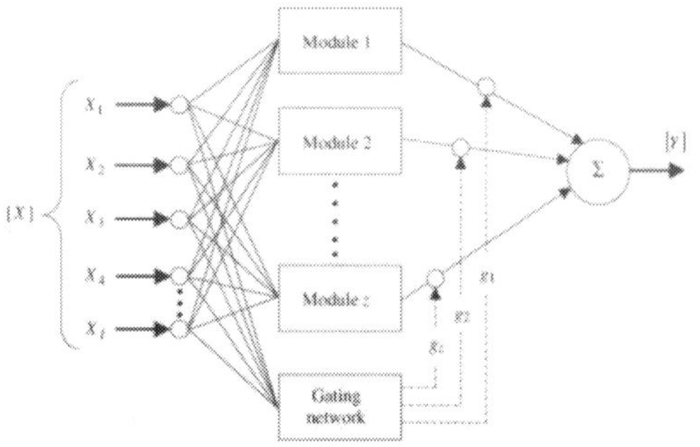

Estimation of:

Injuries.

Fatalities.

Losses in equipment and infrastructure.

Affectation to nearby communities.

Excel, Python, R.

2023 GLOBAL CONFERENCE ON PROCESS SAFETY & BIG DATA
CCPS EPSC
OCTOBER 24 - 25, 2023
DECHEMA HOUSE | Frankfurt am Main, Germany

Monte Carlo Simulation → Mathematical Modelling → Estimation of loses after process safety events

Neural Network to correlate variables:

```
if __name__ == "__main__":

    #initializing the neuron class
    neural_network = NeuralNetwork()

    print("Beginning Randomly Generated Weights: ")
    print(neural_network.synaptic_weights)

    #training data consisting of 4 examples--4 input values and 2 output
    training_inputs = np.array([[0,0,1,0],
                                [1,1,1,1],
                                [1,0,1,0],
                                [0,1,1,1]])

    training_outputs = np.array([[0,1,1,0],[1,0,0,1]]).T

    #training taking place
    neural_network.train(training_inputs, training_outputs, 15000)

    print("Ending Weights After Training: ")
    print(neural_network.synaptic_weights)

# To test or estimate potential outputs from the model

print(neural_network.think(np.array([1,1,0,0])))
```

How to feed a Neural Network using Process Safety Data?:

IChemE
Safety & Loss Prevention Special Interest Group

Sector		Date	Event	Country	Type	Fatalities	Injuries	Abnormal Operations (Context)	Escalation Potential (Context)
Oil & Gas	Upstream	06-Jul-88	Piper Alpha	UK	Explosion	167	[illegible]		
Oil & Gas	Upstream	27-Jul-05	Mumbai High North	India	Explosion	22	[illegible]		
Oil & Gas	Upstream	20-Apr-10	Macondo	USA	Explosion	11	17	X	
Oil & Gas	Upstream	11-Feb-15	Camarupim	Brazil	Explosion	9	26	X	
Oil & Gas	LNG	19-Jan-04	Skikda	Algeria	Explosion	27	74		X
Oil & Gas	OFP	25-Sep-98	Longford	Australia	Fire	2	8	X	
Oil & Gas	Ship	24-Mar-89	Valdez	USA	Pollution	0	0		X
Oil & Gas	Rail	06-Jul-13	Lac Mégantic	Canada	Fire	47	[illegible]		
Oil & Gas	Downstream	04-Jan-66	Feyzin	France	BLEVE	18	84		
Oil & Gas	Downstream	23-Jul-84	Romeoville	USA	Explosion	17	22		
Oil & Gas	Downstream	22-Mar-87	Grangemouth	UK	Explosion	1	0	X	X
Oil & Gas	Downstream	24-Jul-94	Milford Haven	UK	Explosion	0	26	X	X
Oil & Gas	Downstream	23-Feb-99	Avon	USA	Fire	4	1	X	
Oil & Gas	Downstream	17-Aug-99	Izmit	Turkey	Fire	0	0		X
Oil & Gas	Downstream	16-Apr-01	Humber	UK	Explosion	0	5		X
Oil & Gas	Downstream	23-Mar-05	Texas City	USA	Explosion	15	180	X	
Oil & Gas	Downstream	05-Nov-05	Delaware City	USA	Asphyxiation	2	0		
Oil & Gas	Downstream	16-Feb-07	McKee	USA	Fire	0	4		X
Oil & Gas	Downstream	02-Apr-10	Anacortes	USA	Explosion	7	0	X	
Oil & Gas	Downstream	11-Mar-11	Chiba	Japan	BLEVE	0	6		X
Oil & Gas	Downstream	06-Aug-12	Richmond	USA	Fire	0	26		
Oil & Gas	Downstream	18-Feb-15	Torrance	USA	Explosion	0	4	X	X
Oil & Gas	Terminal	19-Nov-84	San Juan Ixhuatepec	Mexico	BLEVE	542	4,248		
Oil & Gas	Terminal	11-Dec-05	Buncefield	UK	Explosion	0	43		X
Oil & Gas	Terminal	23-Oct-09	Bayamón	Puerto Rico	Explosion	0	3		X

Further column groups: "Natech" Triggers / Natural Hazards, Design Factors, Operations Factors.

Source: IChemE Learning Lessons From Major Incidents.

Co-organized by CCPS & EPSC

2023 GLOBAL CONFERENCE ON PROCESS SAFETY & BIG DATA
CCPS EPSC
OCTOBER 24 - 25, 2023
DECHEMA HOUSE | Frankfurt am Main, Germany

Source of Information

Around 770 major accident cases:

- Chemical Safety and Hazard Investigation Board (CSB-US).
- European Major Accident Reporting System (EMARS).
- Failure Knowledge Database (FKD-Japan).
- Accident Reporting Information Analysis (ARIA).
- MARSH Database: The 100 Largest Loses.
- IChemE.

Mathematical Model to Estimate Potential Damages of Process Safety Events on Oil and Gas Facilities.

Process Safety Event:	Moderate Explosion
Type of Process:	Terminals
SIS Level:	SIL I
Building Risk Assessment:	Moderate conditions of integrity.
Nearby communities:	Community less than 500m from Facility.
Business Continuity:	Low Impact in the Business.

PSM Event:	0.7	1
		Factor Neural Network
# Direct Workers:	100	0.9998

VARIABLE:	VALUE	Factor Neural Network
# Contractors exposed:	450	0.9978
OME CCR Operator Shelters:	15	0.9897
Type of Process:	0.8	0.9996
Factor of Congestion:	0.7	0.9998
SIS Level:	1	0.9998

ESTIMATED OUTPUT:	Factor:	Consequences:
Fatalities Direct Workers:	0.185183	18.52
Injuries in Direct Workers:	0.314748	31.47
Fatalities in Contractors:	0.066519	29.93
Injuries in Contractors:	0.332596	149.67
OME OPE Buildings Affected:	0.41237	6.19
Affectation Communities:	0.185072	19%
Loses in M US$	0.185183	185.18
Affectation to Process:	0.493722	49%

2023 GLOBAL CONFERENCE ON PROCESS SAFETY & BIG DATA
CCPS EPSC
OCTOBER 24 - 25, 2023
DECHEMA HOUSE | Frankfurt am Main, Germany

Advantages:

- Easy to use once the Neural Network is trained.
- Compatibility PYTHON, R, MATLAB.
- Applicable also to NATECH cases.
- It combine qualitative and quantitative data.

Limitations:

- No too much data available.
- The model cannot be generalized: Specific to be effective.
- The Model must be customized.

Co-organized by CCPS & EPSC

2023 GLOBAL CONFERENCE ON PROCESS SAFETY & BIG DATA

CCPS EPSC

OCTOBER 24 - 25, 2023
DECHEMA HOUSE | Frankfurt am Main, Germany

How the proposed model can be useful?:

- To identify potential credible scenarios in oil and gas facilities.
- To determine needs and resources to respond effectively to the emergencies in terms of people, equipment , transportation and communication.
- Identification of training needs during emergency drills.
- To facilitate decision-making process during actual emergencies.
- To evaluate effectiveness of Pre-Incident Plans.

July 2019 – LAGOS Nigeria Pipeline explosion.

2023 GLOBAL CONFERENCE ON PROCESS SAFETY & BIG DATA

CCPS EPSC

OCTOBER 24 - 25, 2023
DECHEMA HOUSE | Frankfurt am Main, Germany

2023 GLOBAL CONFERENCE ON PROCESS SAFETY & BIG DATA

CCPS EPSC

2023 GLOBAL CONFERENCE ON PROCESS SAFETY & BIG DATA

CCPS EPSC

University of Sheffield

Martha Scattergood
Louis Allen
Joan Cordiner

Department of Chemical and Biological Engineering

24/10/2023

Health and Safety in Preventive Maintenance

Implementing risk assessment tools and safety regulations into a framework for maintenance optimization

Co-organized by CCPS & EPSC

2023 GLOBAL CONFERENCE ON PROCESS SAFETY & BIG DATA

CCPS EPSC

OCTOBER 24 - 25, 2023
DECHEMA HOUSE | Frankfurt am Main, Germany

Why is preventive maintenance important?

University of Sheffield

- **Texas City Refinery explosion** - 2005, 15 fatalities
- **Cause of explosion**: machine failure due to lack of maintenance schedule
- Regulations related to maintenance planning **were not complied with**
- Implementing an efficient maintenance schedule **increases preventive maintenance** and reduces failures

Co-organized by CCPS & EPSC

2023 GLOBAL CONFERENCE ON PROCESS SAFETY & BIG DATA

CCPS EPSC

OCTOBER 24 - 25, 2023
DECHEMA HOUSE | Frankfurt am Main, Germany

What literature shows

University of Sheffield

Combination of:

- **Statistical tools**: Bayesian networks, Fault Tree Analysis, Markov Models, Machine Learning
- **Health and safety tools**: HAZOP, FMEA

To create a maintenance schedule applied to **High-risk units only,** health and safety regulations are not taken into account.

Example – HAZOP-DBN model by Hu et. al[1]

[1] Hu, J., Zhang, L., and Liang, W. (2012). Opportunistic predictive maintenance for complex multi-component systems based on dbn-hazop model. Process Safety and Environmental Protection. 90:376–388.

Co-organized by CCPS & EPSC

How this has been applied to preventive maintenance

University of Sheffield

Combination of:

- **Statistical tools**: Bayesian networks, Fault Tree Analysis, Markov Models, Machine Learning
- **Health and safety tools**: HAZOP, FMEA

To create a maintenance schedule applied to **High-risk units only,** health and safety regulations are not taken into account.

Example – HAZOP-DBN model by Hu et. al.[1]

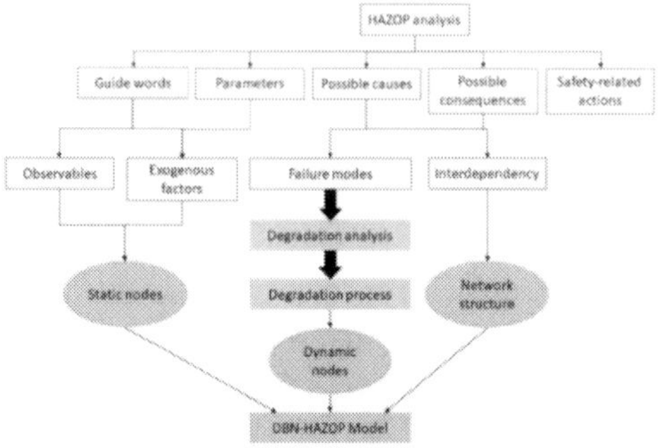

[1] Hu, J., Zhang, L., and Liang, W. (2012). Opportunistic predictive maintenance for complex multi-component systems based on dbn-hazop model. Process Safety and Environmental Protection, 90:376–388

Co-organized by CCPS & EPSC

2023 GLOBAL CONFERENCE ON PROCESS SAFETY & BIG DATA
CCPS EPSC
OCTOBER 24 - 25, 2023
DECHEMA HOUSE | Frankfurt am Main, Germany

Literature shows a gap regarding key health and safety requirements

University of Sheffield

- Regulations have not been implemented into a maintenance scheduling framework
- Safety tools combined with statistical tools are only used for high-risk equipment
- Among these safety tools, LOPA is overlooked

Co-organized by CCPS & EPSC

2023 GLOBAL CONFERENCE ON PROCESS SAFETY & BIG DATA

CCPS EPSC

OCTOBER 24 - 25, 2023
DECHEMA HOUSE I Frankfurt am Main, Germany

Filling the gap to optimize preventive maintenance planning

University of Sheffield

HAZOP and **LOPA** combined with a **Hidden Markov Model** to determine the ideal frequency of preventive maintenance for ALL equipment.

Key **rules and regulations** implemented into a flowsheet to determine the ideal frequency of preventive maintenance for equipment associated with these regulations.

Framework assessing ALL equipment and resulting in **full compliance with health and safety regulations**.

Co-organized by CCPS & EPSC

2023 GLOBAL CONFERENCE ON PROCESS SAFETY & BIG DATA

CCPS EPSC

OCTOBER 24 - 25, 2023
DECHEMA HOUSE I Frankfurt am Main, Germany

Using a Hidden Markov Model (HMM) to predict failure and consequence probabilities

University of Sheffield

$$P(Z_{1:n}, X_{1:n}) = P(Z_1)P(X_1|Z_1)\prod_{t=2}^{n} P(Z_t|Z_{t-1})P(X_t|Z_t)$$

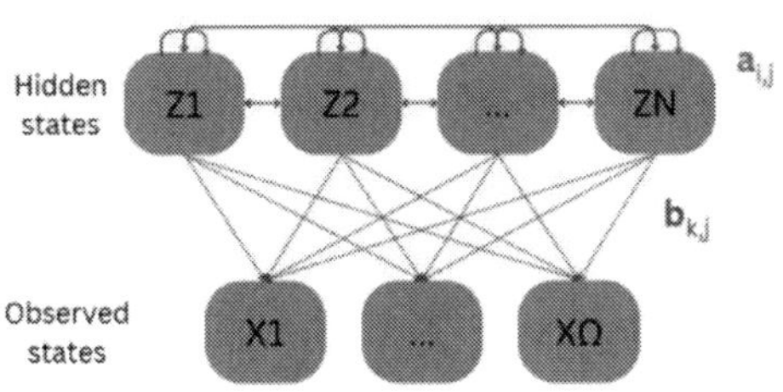

$a_{i,j}$= hidden state transition probability

$b_{k,j}$= hidden state to observable state transition probability

- The Hidden Markov Model uses past data to predict machine failure
- The hidden variables (Z) represent the failures and the observed variables (X) represent the consequences

Co-organized by CCPS & EPSC

2023 GLOBAL CONFERENCE ON PROCESS SAFETY & BIG DATA

CCPS EPSC

OCTOBER 24 - 25, 2023
DECHEMA HOUSE | Frankfurt am Main, Germany

Determining the urgency of a failure using HMM and HAZOP/LOPA

University of Sheffield

Guide word	Deviation	Possible causes	Prob.	Consequences	Prob.	LOPA - Layers
HIGH	1. High Pressure	1.1 Obstruction in outlet pipe causing a build up of pressure 1.2. Imbalance of flow (higher flow in than out) causing a build up of pressure inside the reactor 1.3. Fouling in the sieve trays slowing the flow inside the reactor and hence increasing the pressure below the fouled sieve tray	[illegible] [illegible] [illegible]	1.a. Rupture in reactor or pipes 1.b. Explosion 1.c. Fire 1.d. Release of toxic gases 1.e. Spillage of toxic liquids	[illegible] [illegible] [illegible] [illegible] [illegible]	1. Process Reactor design to not exceed 150 bar. 2. BPCS Control valve in inlet stream linked to pressure indicator 3. AOI Pressure indicators and alarms at the bottom and top of the reactor to indicate any pressure build up. Flow indicators and alarms in the inlet and outlet pipes to indicate any imbalances. 4. SIS Trip system closing inlet control valve 5. Physical Protection Pressure relief valve 6. Physical containment Bunding around reactor

Calculated from past data using HMM → P_{Zi}

Taken from 5x5 risk matrix → Sev_{Xi}, L_{Xi}

$$RN = \frac{P_{Zi}\sum_{i=1}^{n} Sev_{Xi}\cdot L_{Xi}}{n}$$

Taken from HAZOP/LOPA → n

P_{Zi} = Probability of concerned failure Zi

Sev_{Xi} = Severity of the consequences of Xi

L_{Xi} = Likelihood of the consequences of Xi

n = number of consequences associated with Zi

Co-organized by CCPS & EPSC

2023 GLOBAL CONFERENCE ON PROCESS SAFETY & BIG DATA

CCPS EPSC

OCTOBER 24 - 25, 2023
DECHEMA HOUSE | Frankfurt am Main, Germany

Determining the urgency of a failure using HMM and HAZOP/LOPA

University of Sheffield

Calculated from past data using HMM → P_{Zi}

Taken from 5x5 risk matrix → Sev_{Xi}, L_{Xi}

$$RN = \frac{P_{Zi}\sum_{i=1}^{n} Sev_{Xi}\cdot L_{Xi}}{n}$$

Taken from HAZOP/LOPA → n

P_{Zi} = Probability of concerned failure Zi

Sev_{Xi} = Severity of the consequences of Xi

L_{Xi} = Likelihood of the consequences of Xi

n = number of consequences associated with Zi

SEVERITY

LIKELIHOOD	1	2	3	4	5
1	1/D	2/D	3/D	4/D	5/D
2	2/D	4/D	6/D	8/D	10/D
3	3/D	6/D	1/D	12/D	15/D
4	4/D	8/D	12/D	16/D	20/D
5	5/D	10/D	15/D	20/D	25/D

- Once a year
- Once every 2 years
- Once every 4 years
- Once every 10 years

D = number of deviations in HAZOP

Co-organized by CCPS & EPSC

Incorporating Health and Safety regulations

University of Sheffield

Regulations taken into account for the UK are:

- The International Society of Automation (ISA)
- The Control of Major Accident Hazards (COMAH)
- Dangerous Substance and Explosive Atmospheres (DSEAR)
- Public Protection Regulations
- Pressure Systems Safety Regulations (PSSR)

2023 GLOBAL CONFERENCE
ON PROCESS SAFETY & BIG DATA
CCPS EPSC
OCTOBER 24 - 25, 2023
DECHEMA HOUSE | Frankfurt am Main, Germany

The final framework

University of Sheffield

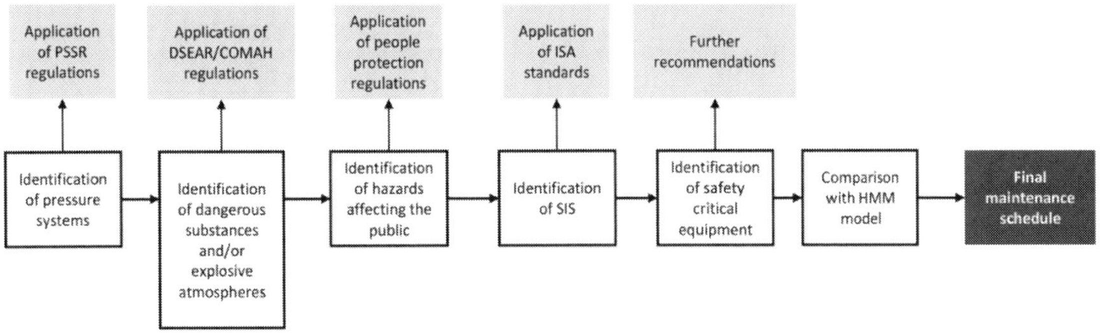

Co-organized by CCPS & EPSC

Case study application - urea production plant

University of Sheffield

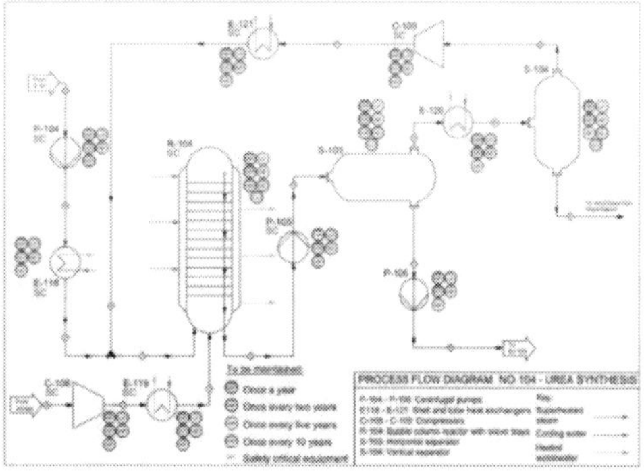

- "Past data" was simulated by generating 1120 sequences of 4 deviations
- Using the HMM model the deviation probabilities were calculated
- A HAZOP/LOPA document was created using these deviations
- All units shown on the PFD were then put through the regulations flowsheet

2023 GLOBAL CONFERENCE ON PROCESS SAFETY & BIG DATA
CCPS EPSC
OCTOBER 24 - 25, 2023
DECHEMA HOUSE | Frankfurt am Main, Germany

Case study application - urea production plant

University of Sheffield

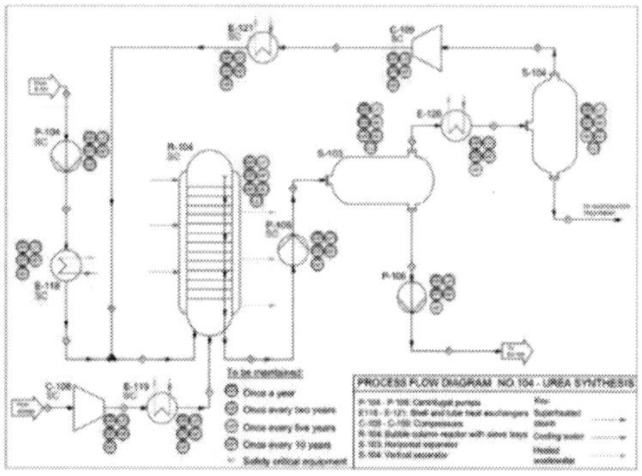

- Simulated data
- Initial deviation probability calculation
- HMM deviation probability calculation
- Risk number calculation
- HMM preventive maintenance schedule
- Regulations flowsheet
- Comparison with HMM
- **Final schedule**

Co-organized by CCPS & EPSC

What the results show

University of Sheffield

- For safety critical systems, the regulations flowsheet has the **higher or equal maintenance frequency**
- For non-safety critical systems, **regulations apply much less**
- The regulations flowsheet also serves as a basis for **inspections, proof testing and timings for maintenance scheduling**

Co-organized by CCPS & EPSC

2023 GLOBAL CONFERENCE ON PROCESS SAFETY & BIG DATA
CCPS EPSC
OCTOBER 24 - 25, 2023
DECHEMA HOUSE | Frankfurt am Main, Germany

Final conclusions

University of Sheffield

- Safety critical and other high-risk equipment complies with health and safety regulations, ensuring they **stay within the law**
- Equipment perceived as low risk that are often overlooked will be assessed with the help of past data and scheduled accordingly, **no single unit is overlooked or ignored altogether**
- A framework such as this would help to **avoid future disasters** such as Texas City Refinery

Co-organized by CCPS & EPSC

2023 GLOBAL CONFERENCE ON PROCESS SAFETY & BIG DATA
CCPS EPSC

OCTOBER 24 - 25, 2023
DECHEMA HOUSE | Frankfurt am Main, Germany

Further action to improve preventive maintenance

University of Sheffield

- The case study used involved simulated "past data"
  - **Do you have a real-life case study that we could work on**?
- The model could be improved by integrating neighbouring units together, a deviation in one unit could cause a deviation in the subsequent unit
- Regulations will differ between countries and even companies
- Maintenance history, age, size and materials of the unit with the use of additional unit characterisation methods should be integrated into the framework

Co-organized by CCPS & EPSC

2023 GLOBAL CONFERENCE ON PROCESS SAFETY & BIG DATA
CCPS EPSC

OCTOBER 24 - 25, 2023
DECHEMA HOUSE | Frankfurt am Main, Germany

University of Sheffield

Questions?

For any additional questions, please contact Martha Scattergood
martha.scattergood@loreal.com
or Louis Allen
lallen6@sheffield.ac.uk

Co-organized by CCPS & EPSC

2023 GLOBAL CONFERENCE
ON PROCESS SAFETY & BIG DATA

CCPS EPSC

OCTOBER 24 - 25, 2023
DECHEMA HOUSE | Frankfurt am Main, Germany

Spatiotemporal Fault Detection and Diagnosis (FDD) to Enhance Process Safety

The University of Sheffield

Louis Allen, Prof. Haiping Lu, Prof. Joan Cordiner

Funding from the University of Sheffield & EPSRC

University of Sheffield

Co-organized by CCPS & EPSC

2023 GLOBAL CONFERENCE
ON PROCESS SAFETY & BIG DATA

CCPS EPSC

OCTOBER 24 - 25, 2023
DECHEMA HOUSE | Frankfurt am Main, Germany

Research Background

- The benefits of FDD are multi-faceted:
  - Safer operation
  - Increased sustainability
  - Reduced costs

- FDD using machine learning has become a popular field in the wake of I4.0

University of Sheffield

2023 GLOBAL CONFERENCE ON PROCESS SAFETY & BIG DATA

CCPS EPSC

OCTOBER 24 - 25, 2023
DECHEMA HOUSE | Frankfurt am Main, Germany

FDD Technologies in Process Manufacturing

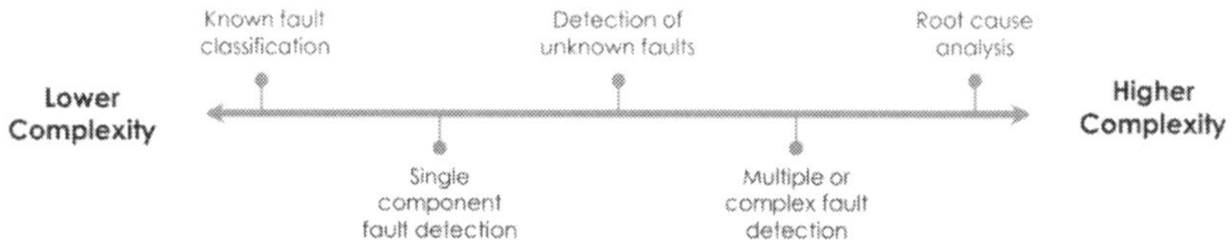

University of Sheffield

2023 GLOBAL CONFERENCE ON PROCESS SAFETY & BIG DATA

CCPS EPSC

OCTOBER 24 - 25, 2023
DECHEMA HOUSE | Frankfurt am Main, Germany

FDD Technologies in Process Manufacturing

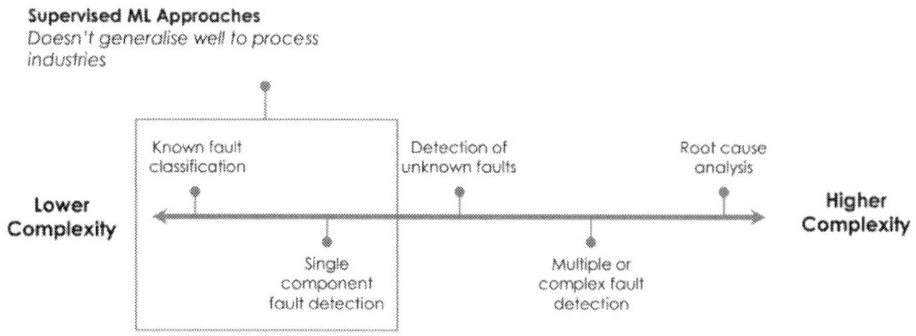

University of Sheffield

2023 GLOBAL CONFERENCE ON PROCESS SAFETY & BIG DATA

CCPS EPSC

OCTOBER 24 - 25, 2023
DECHEMA HOUSE | Frankfurt am Main, Germany

FDD Technologies in Process Manufacturing

Largely human based processes

- *Time consuming*
- *High expertise requirement*

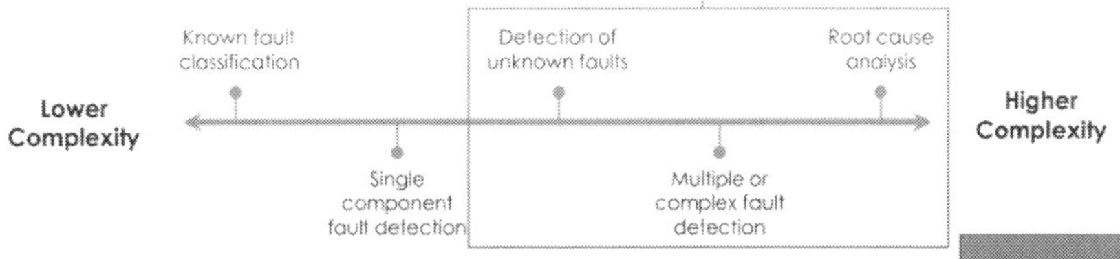

How can these processes be automated?

University of Sheffield

2023 GLOBAL CONFERENCE ON PROCESS SAFETY & BIG DATA

CCPS EPSC

OCTOBER 24 - 25, 2023
DECHEMA HOUSE | Frankfurt am Main, Germany

Combining domain knowledge with AI

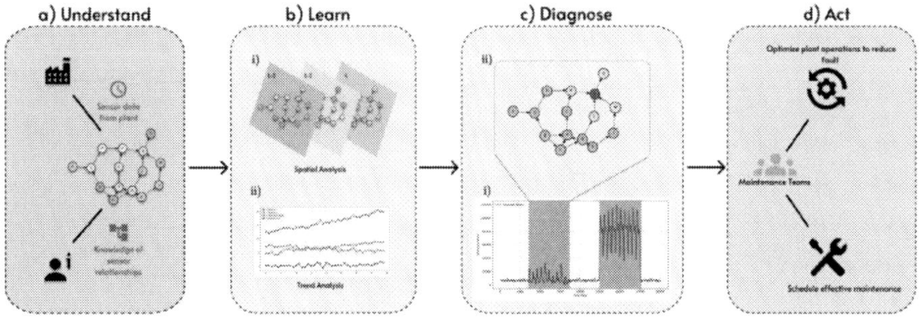

University of Sheffield

2023 GLOBAL CONFERENCE ON PROCESS SAFETY & BIG DATA

CCPS EPSC

OCTOBER 24 - 25, 2023
DECHEMA HOUSE | Frankfurt am Main, Germany

How can we integrate domain knowledge?

We use **mechanistic equations** to build **knowledge graphs** of processes

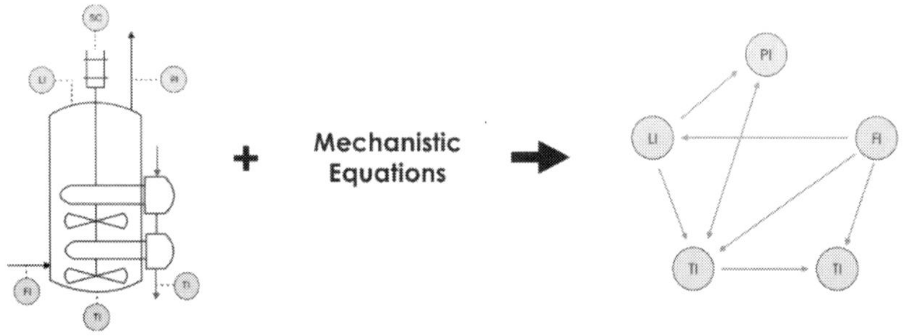

University of Sheffield

2023 GLOBAL CONFERENCE ON PROCESS SAFETY & BIG DATA

CCPS EPSC

OCTOBER 24 - 25, 2023
DECHEMA HOUSE | Frankfurt am Main, Germany

Tennessee Eastman Process

Produces 2 products (G, H) from 4 reactants (A, C, D, E) with 2 by-products (B, F).

5 major unit operations:
- **R003 Reactor**
- **C115 Condenser**
- E005 Stripper
- **S102 Vapor-Liquid Separator**
- **K100 Recycle Compressor**

41 measured variables

11 manipulated variables

15 simulated fault scenarios

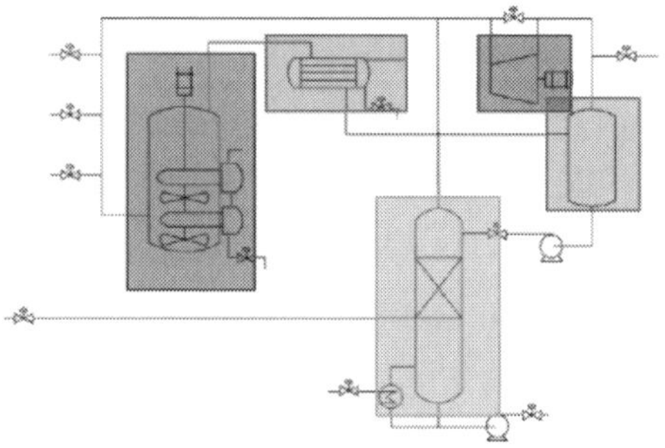

Process and instrumentation diagram of the Tennessee Eastman Process[1]. Major units highlighted.

University of Sheffield

[1] Downs, J.J. and Vogel, E.F., 1993. A plant-wide industrial process control problem. Computers & chemical engineering, 17(3), pp.245-255

Tennessee Eastman Process

TEP as a knowledge graph:

$$G = (V, E, \mathbf{W}),$$

$$|V| = n,$$

$$e_{i,j} \in E,$$

$$\mathbf{W}_{i,j} = \begin{cases} \exp\left(-\frac{\mathrm{dist}(v_i, v_j)^2}{\sigma^2}\right), & i \neq j, e_{i,j} = 1 \\ 0, & \text{otherwise,} \end{cases}$$

Given a process graph, G, and sensor data $\mathbf{X} \in \mathbb{R}^{m \times n}$ how can we learn spatiotemporal correlations to detect anomaly?

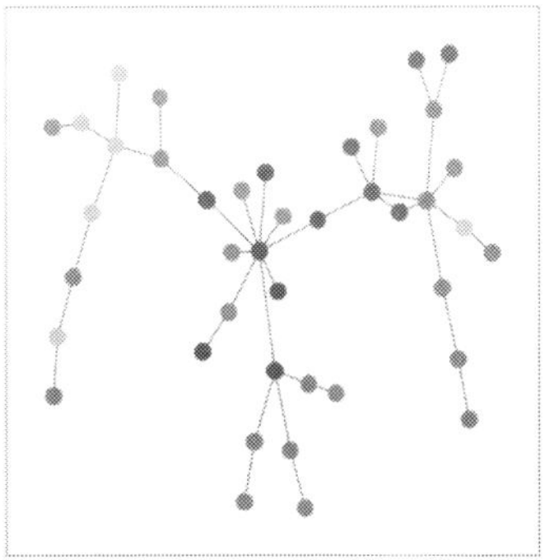

Tennessee Eastman process as a knowledge graph based on mechanistic equations.

2023 GLOBAL CONFERENCE ON PROCESS SAFETY & BIG DATA
CCPS EPSC
OCTOBER 24 - 25, 2023
DECHEMA HOUSE | Frankfurt am Main, Germany

How do we use the knowledge graph?

Architecture based on a Generative Adversarial Network:

- Generator attempts to fool the discriminator.
- Discriminator tries to tell generated sequence vs real.

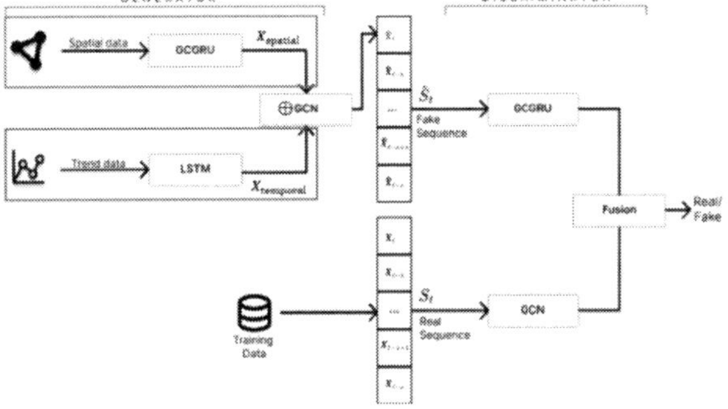

University of Sheffield

Assess the spatial information

- Look for **contextual anomaly.**
- The dynamics of a sensor can be described by the k nearest sensors.

$$G_v = (V_v, E_v, \boldsymbol{W}_v)$$

- Predict the dynamics of each node using a GCGRU across time period s – **based on the subgraph of the node.**

$$\boldsymbol{X}_{\text{spatial},v} = \left[\boldsymbol{X}_{v,t-s}, \boldsymbol{X}_{v,t-s+1}, \dots, \boldsymbol{X}_{v,t-1}\right] \in \mathbb{R}^{s \times k}$$

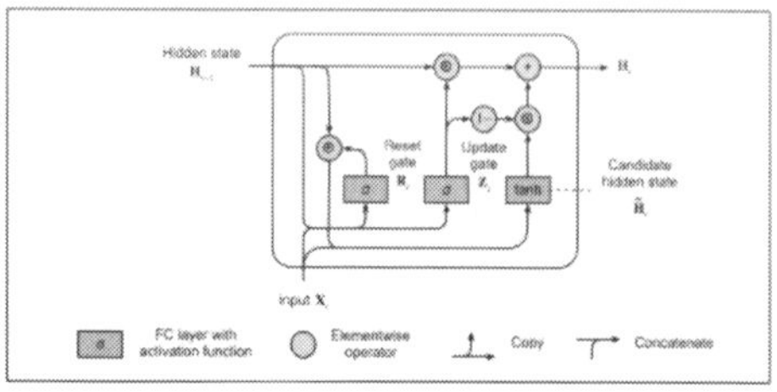

$$\boldsymbol{r}^{(t)} = \sigma(\boldsymbol{g}_{\theta,r} *_G \left[\boldsymbol{X}^{(t)}, \boldsymbol{H}^{t-1}\right] + \boldsymbol{b}_r) \quad \boldsymbol{C}^{(t)} = \tanh(\boldsymbol{g}_{\theta,C} *_G \left[\boldsymbol{X}^{(t)}, (\boldsymbol{r}^{(t)} \odot \boldsymbol{H}^{t-1})\right] + \boldsymbol{b}_c)$$

$$\boldsymbol{u}^{(t)} = \sigma(\boldsymbol{g}_{\theta,u} *_G \left[\boldsymbol{X}^{(t)}, \boldsymbol{H}^{t-1}\right] + \boldsymbol{b}_u) \quad \boldsymbol{H}^{(t)} = \boldsymbol{u}^{(t)} H^{(t-1)} + \left(1 - \boldsymbol{u}^{(t)}\right) \odot \boldsymbol{C}^{(t)}$$

Gated recurrent unit (GRU) architecture with modified equations to use graph convolutions[1].

University of Sheffield

[1] Seo, Y., Vandergheynst, P., & Bresson, X. (2017). Structured sequence modeling with graph convolutional recurrent networks.

2023 GLOBAL CONFERENCE ON PROCESS SAFETY & BIG DATA
CCPS EPSC
OCTOBER 24 - 25, 2023
DECHEMA HOUSE | Frankfurt am Main, Germany

Assess the temporal information

- Use a long-short-term memory network.
- Predict the **long term degradation** and **cyclical anomaly** over period l.

$$\boldsymbol{x}_{\text{temporal},v} = \left[x_{v,t-l}, x_{v,t-l+1}, \dots, x_{v,t-1}\right] \in \mathbb{R}^{l},$$

- Combine with spatial prediction to give our overall forecast

$$\widehat{\boldsymbol{X}}_t = \sigma\left(g_\theta *_G \left[\boldsymbol{X}_{\text{spatial}}, \boldsymbol{x}_{\text{temporal}}\right]\right)$$

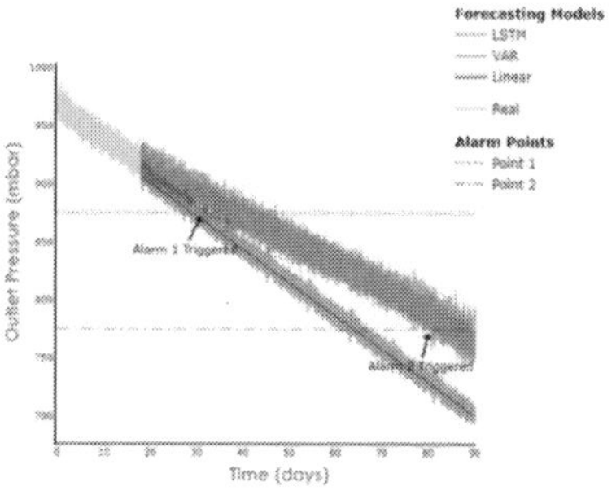

Forecasting inverse-sigmoidal degradation using machine learning models. LSTM proved most accurate architecture.[1]

University of Sheffield

[1] Ahmed, M., Allen, L. and Cordiner, J., 2022, November. Investigating Machine Learning Techniques for Effective Predictive Maintenance in Industrial Systems. In 2022 AIChE Annual Meeting. AIChE.

Score based fault detection

- Use the prediction to calculate a loss from both discriminator and generator.
- Combine to create an anomaly score for each node at each time-step.

$$s_G = \left\| G_\theta(v,t) - X_{v,t} \right\|_2$$

$$s_D = D_\phi(X_{v,t}) - D_\phi(\hat{X}_{v,t})$$

$$anom(v,t) = s_G - \frac{s_D}{\lambda_s}$$

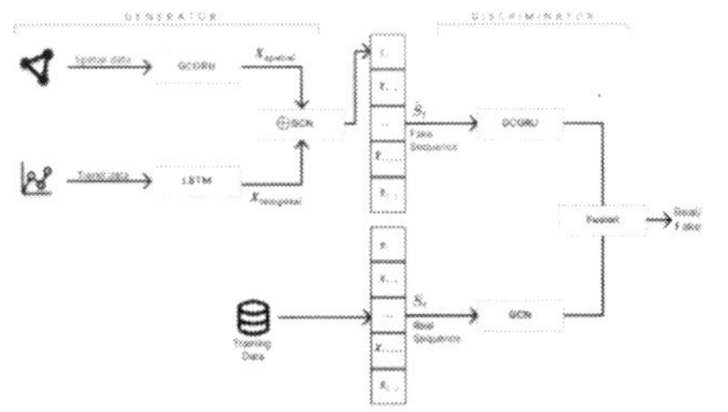

University of Sheffield

2023 GLOBAL CONFERENCE ON PROCESS SAFETY & BIG DATA
CCPS EPSC
OCTOBER 24 - 25, 2023
DECHEMA HOUSE | Frankfurt am Main, Germany

High performance compared to SOA

Key results:

- **Fault Detection Rate – 96.2%**
- **F1 Score – 0.979**

- Maintained a higher average fault detection ability and F1 score than state-of-the-art
- Achieved high performance on 'challenging' fault cases.

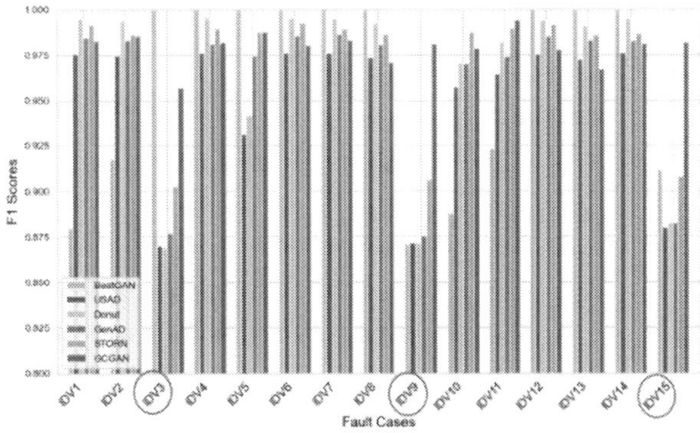

University of Sheffield

2023 GLOBAL CONFERENCE ON PROCESS SAFETY & BIG DATA

CCPS EPSC

OCTOBER 24 - 25, 2023
DECHEMA HOUSE | Frankfurt am Main, Germany

Fault diagnosis (RCA)

- Look for a change in the anomaly scores of **node subgraphs** before and after the detected fault.
- Shows which nodes are **most likely to be contributing** to the fault.

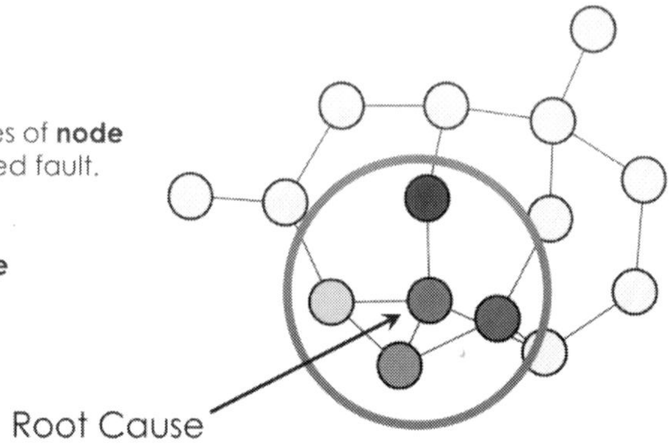

2023 GLOBAL CONFERENCE ON PROCESS SAFETY & BIG DATA

CCPS EPSC

OCTOBER 24 - 25, 2023
DECHEMA HOUSE | Frankfurt am Main, Germany

Fault diagnosis (RCA)

- Identified true root cause (red) as primary contributor in **65% of cases with no prior knowledge**.
- Identified nodes on fault propagation (orange) pathway in **85% of cases tested**.

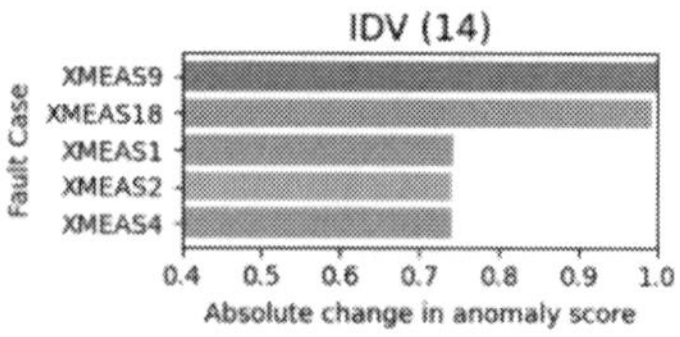

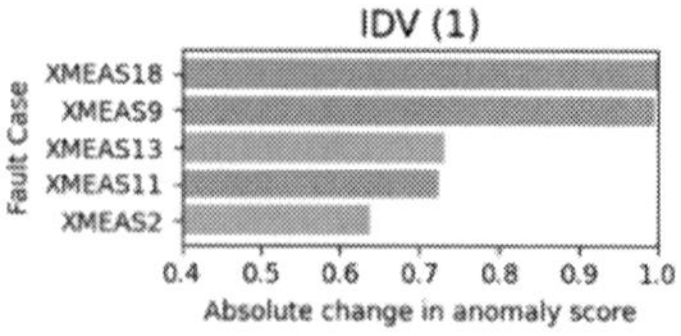

2023 GLOBAL CONFERENCE ON PROCESS SAFETY & BIG DATA

CCPS EPSC

OCTOBER 24 - 25, 2023
DECHEMA HOUSE | Frankfurt am Main, Germany

Conclusion

- Presented a method that can integrate domain knowledge into ML models for fault detection and diagnosis.
- Results show improvement over state-of-the-art in fault detection ability.
- We demonstrate robustness by maintaining high performance in 'difficult' fault cases.
- Showed that integrating knowledge allows the model to diagnose faults to find true root cause.
- The impact is a significant reduction in the time between the fault occurrence, and effective intervention

University of Sheffield

Co-organized by CCPS & EPSC

2023 GLOBAL CONFERENCE ON PROCESS SAFETY & BIG DATA

CCPS EPSC

OCTOBER 24 - 25, 2023
DECHEMA HOUSE | Frankfurt am Main, Germany

Thank you for your attention.

Funding From

EPSRC

University of Sheffield

Any questions?

lallen6@sheffield.ac.uk

Co-organized by CCPS & EPSC

We are offering early access to this work. **Scan below for more details:**

2023 GLOBAL CONFERENCE ON PROCESS SAFETY & BIG DATA
CCPS EPSC

OCTOBER 24 - 25, 2023
DECHEMA HOUSE | Frankfurt am Main, Germany

Tools to Avoid Human Errors: Attack operator performance issues from all directions

David A Strobhar, PE
Beville/Center for Operator Performance

Co-organized by CCPS & EPSC

2023 GLOBAL CONFERENCE ON PROCESS SAFETY & BIG DATA
CCPS EPSC

OCTOBER 24 - 25, 2023
DECHEMA HOUSE | Frankfurt am Main, Germany

Roots of human factors

Speed/Capability

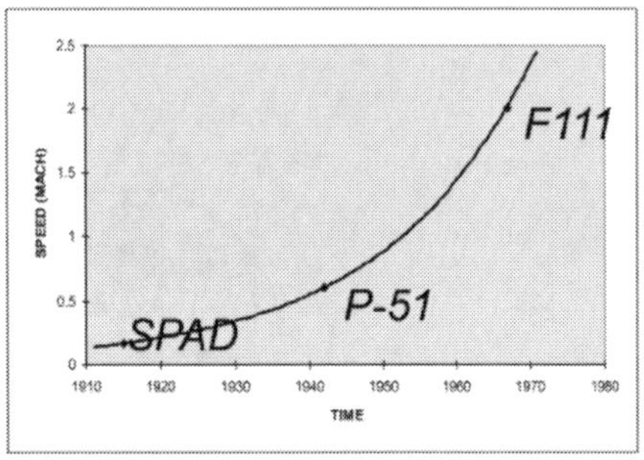

Number of Controls/Switches

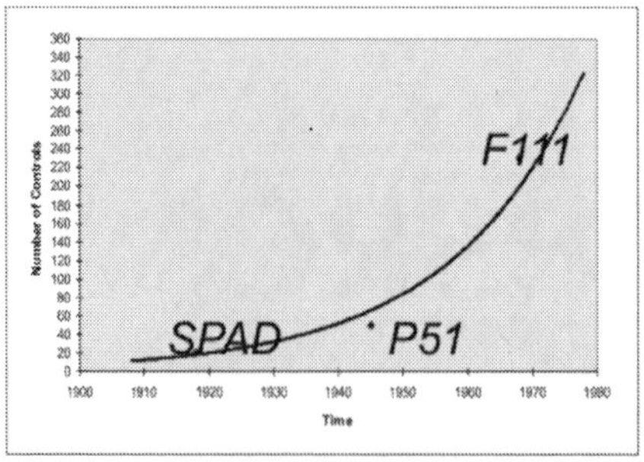

Nuclear Power

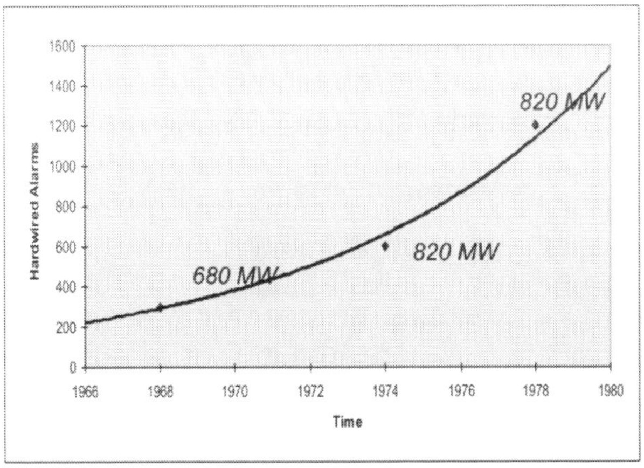

Performance Shaping Factors

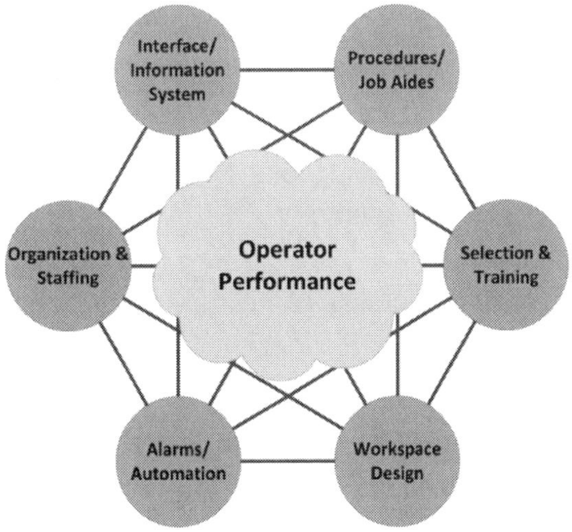

4

Center for Operator Performance

www.operatorperformance.net

2023 GLOBAL CONFERENCE ON PROCESS SAFETY & BIG DATA

CCPS EPSC

OCTOBER 24 - 25, 2023
DECHEMA HOUSE | Frankfurt am Main, Germany

Alarms/Automation

- Multi-Phase Project
  - Age induced hearing loss
  - Larger control rooms
- What doesn't help
  - Increasing loudness (Amplitude modulation)
  - Making the sound longer
- Developed 14 alarm sets
  - Convey urgency
  - Learnable
  - Localizable

Co-organized by CCPS & EPSC

It's not the volume, but the frequency

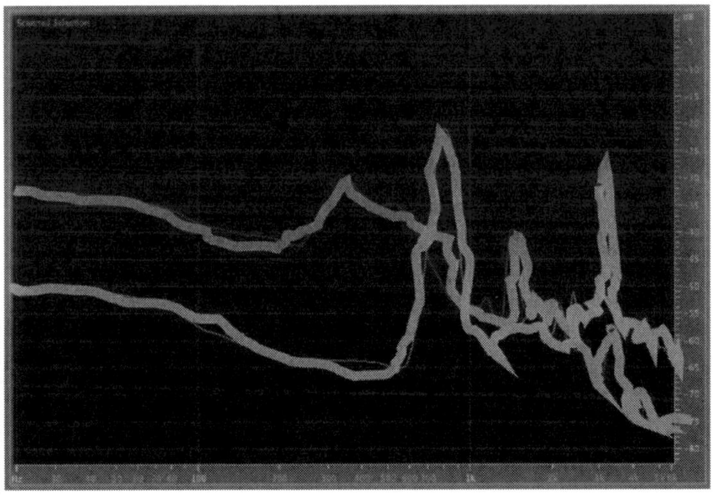

Red = Ambient;

Green = Alarm sound

7

sensitivities to certain

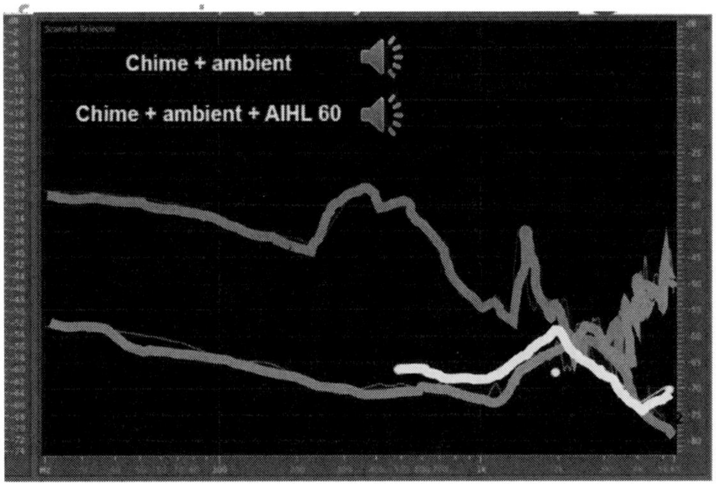

Red = Ambient;

Green = Alarm sound

Yellow - AIHL

8

Interface

- Data to Information Handbook: Refining Edition
- Display Design Scorecard

Co-organized by CCPS & EPSC

2023 GLOBAL CONFERENCE ON PROCESS SAFETY & BIG DATA
CCPS EPSC

Using this document

- Ideas to improve existing graphics
- Create display hierarchy for entire unit

Key Variables

Level	Category	Parameter	Tag	Description
L1	Common	O_2 in Flue Gas	AC XX41	
		NOx in Flue Gas	AI XX41	NOx may be measured, if it is it will be a critical parameter
		Outlet Temperature	T XX38	
		Process Fluid Flow	F XX39	Charge rate to the unit
		Fuel Gas	P XX41	
		Draft Pressure	P XX40	
L2	Common	Process Fluid Temperature In	T XX39	
		Process Fluid Temperature Out	TC XX38	
		Process Fluid Outlet Flow	F XX38	
		CO in Flue Gas	AC XX43	If available (also O_2 and CO are highly correlated)
		Fuel Gas Flow	F XX41	
		Air Flow (if measure)	F XX43	
		Individual Pass flow	F XX42	
		Individual Pass Inlet Temperature	T XX42	
		Individual Pass Outlet Temperature	T XX48	If exiting
		Pass Valve Positions	ZV XX42	If available
		Status of Blocking Valve on Fuel Gas	YV XX38	
	Convection section	Stack Temperature (out)	T XX40	
		Convection Temperature (in, below the tubes)	T XX41	
		Material Flow	F XX40	Could be process, steam, boiler feed

Review Grading Criteria

- Grading Global (for whole display system)
  - Hierarchy and Navigation
  - Convention and Coding
- For L1, L2, L3 Displays
  - Supports Key Decisions/Tasks
  - Layout and Display Elements
  - Quantitative/Qualitative
  - Saliency Regions
  - Saliency Distractors

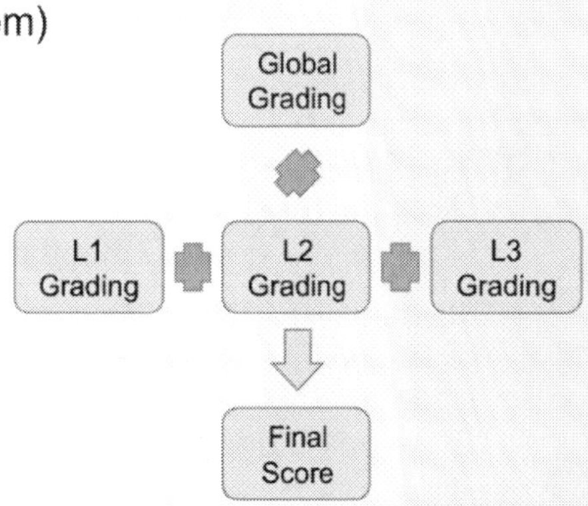

12

Layout and Display Elements Scoring

Criteria	*Score*	*L1 Scoring Guidelines*
Layout and Display Elements	**2.00**	
Organization within individual graphic. L1 should be partitioned with each section mapping to an L2.	4	4- Graphic is partitioned consistent with the L2 and this structure is easily recognized 3- No partitioning but the relationship with L2 structure is still recognized 2- Partitioned but maps to multiple L2's from a single partition 1- Not partitioned and L2 structure not recognizable 0- No cohesive mapping to L2
Display elements should be arranged for ease of scanning and pattern recognition. This should include a logical representation of the system that is consistent with the operator's mental models. Multiple items of the same type of information are placed together (i.e., column temperatures, parallel reactor flows)	1	4- Arranged for easy scanning and pattern recognition; logical representation matching the operator's mental model; like items grouped together. 3- Arranged for easy scanning and pattern recognition; logical representation matching the operator's mental model 2- Logical representation matching the operator's mental model; 1- Does not represent console operator's mental model 0- None of these criteria are met
Contains important information without clutter of low- or no-value information. Static elements (vessels, pumps, piping, labels, etc.) are present to provide context for dynamic elements when useful. Vessel shapes, when used, should enclose related dynamic elements.	1	4- Has no unnecessary static or dynamic elements. Every static element has a strong association with a dynamic element or group of elements. 3- Every static element can be associated with a dynamic element. (e.g., vessel shapes not enclosing dynamic elements) 2- Has some static elements that are not associated with dynamic elements or that are redundent or unnecessary 1- Has several static elements that are not associated with dynamic elements or that are redundent or unnecessary 0- No dynamic information
Dynamic elements are present to support overall situation awareness.	2	4- Every dynamic element can be mapped back to a decision 3- Most dynamic elements can be mapped back to a decision 2- Some dynamic elements can be mapped back to a decision 1- Unclear relationship between dynamic elements and decisions 0- Unnecessary dynamic elements outnumber those that are connected to decisions

13

Training

- Shadowbox/ Decision Making Exercises
  - Decision making is a skill
  - How do you get and maintain skills? PRACTICE!
- Training Benchmarking
  - Variance in operator training; 5-52 weeks for board qualification depending upon company
  - Everyone thought they were producing good operator
- Knowledge Benchmarking

Knowledge Benchmarking

- Evaluate relative benefits of different approaches to training
- Create objective test (measurement tool) on common topic (e.g., distillation)
- Members create questions
  - 50 questions
  - All multiple choice
- Researcher handles statistics
- Validate questions to ensure no ceiling effect
- Can be administered with-in & across plants

2023 GLOBAL CONFERENCE ON PROCESS SAFETY & BIG DATA
OCTOBER 24 - 25, 2023
DECHEMA HOUSE | Frankfurt am

Test E

1. What is an indication of flooding in a distillation tower?
   a) There is an increase in overhead product
   b) There is an increase in bottoms level
   c) There is an increase in product quality
   d) The product quality will remain the same
2. What is the primary concern when a section of your tower is flooded?
   a) There will be an increase in feed from storage
   b) There will be a loss of desired distillation product quality
   c) There will be an increase in steam to the reboiler
   d) There will be high reflux temperature
   e) There will be an increase in overhead temperature
3. How do you bring the tower out of flooding state?
   a) Increase the reflux
   b) Decrease pressure on the tower
   c) Cut the heat to the reboiler
   d) Increase feed to tower

Center for Operator Performance

www.operatorperformance.net

2023 GLOBAL CONFERENCE ON PROCESS SAFETY & BIG DATA

CCPS EPSC

- Center for Operator Performance
  - Public site http://operatorperformance.org/
  - Member site http://operatorperformance.net/
- Beville Engineering Archives
  - http://www.beville.com/
- Human Factors in Process Plant Operation
  - http://www.amazon.com/David-A.-Strobhar/e/B00MSW313G/

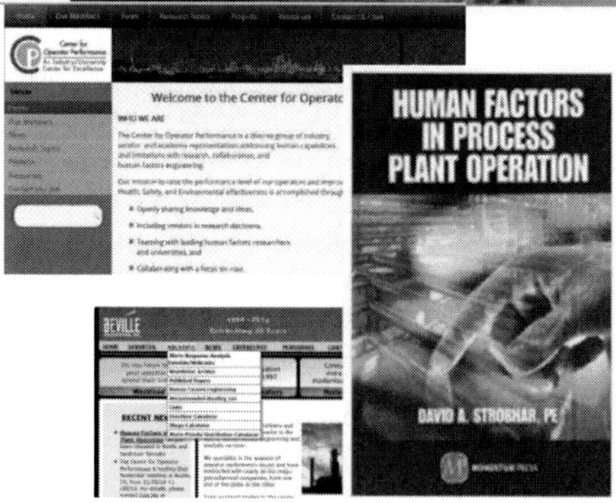

AUTHOR INDEX

AIChE
120 Wall Street, FL 23
New York, NY 10005-4020

ISBN 979-8-3313-3049-1

CURRAN ASSOCIATES INC.
proceedings.com